W0115204

Technology and the Insurance Industry

Antonella Cappiello

Technology and the Insurance Industry

Re-configuring the Competitive Landscape

Antonella Cappiello
Department of Economics and Management
University of Pisa
Pisa, Italy

ISBN 978-3-319-74711-8   ISBN 978-3-319-74712-5   (eBook)
https://doi.org/10.1007/978-3-319-74712-5

Library of Congress Control Number: 2018931930

Cover illustration: Détail de la Tour Eiffel © nemesis2207/Fotolia.co.uk.

Printed on acid-free paper

This Palgrave Pivot imprint is published by Springer Nature
The registered company is Springer International Publishing AG
The registered company address is: Gewerbestrasse 11, 6330 Cham, Switzerland

CONTENTS

LIST OF TABLES

LIST OF BOXES

CHAPTER 1

Introduction

Abstract This chapter introduces the aim of the book and sets its theoretical framework providing a guideline for the topics included in each chapter.

Keywords Insurance industry • InsurTech • Insurance digitalisation • Insurance distribution

Technological innovation is deeply changing the economic and social fabric at all levels, making baseline scenarios changeable and more liquid and asking companies a dynamic and proactive response in order to successfully compete on a global scale.

A great diffusion of technological instruments, able to simulate human behaviours, is also expected in the near future; this will significantly affect organisational and business models.

This will lead to a radical innovation in the ways interaction is carried out with all stakeholders in supply chains, which are becoming more and more diversified and characterised by automatised and shared processes. The competitive scenario is changing, due to the entrance of new operators or operators coming from different sectors into the various markets, thanks to the common digital base. Digital technology deeply impacts insurers' existing business models. This is clearly a challenge, where the key factor is the speed at which insurers will be able to take advantage of the technological innovation, without suffering from its effects.

© The Author(s) 2018
A. Cappiello, *Technology and the Insurance Industry*,
https://doi.org/10.1007/978-3-319-74712-5_1

The growth in the digitalisation process of the insurance sector can be considered as the result of many concurrent factors, which are attributable to the technological development on the one side, both in terms of size and quality of Internet infrastructures and connection opportunities and in terms of usable applications, and attributable, on the other side, to the change in customers' attitude, which sees the entrance of digital natives' generation and an increasing willingness to use virtual channels, especially after having tried them for services other than the insurance service. It is also to note that conjunctural factors have been reducing profits and driving insurers to search for cost reductions, reaching also for digitised processes, be they productive or distributive.

The increasing digitisation of insurance industry raises issues about strategy, risk, market and organisational structure, workforce, and culture—issues that, in the final analysis, require the full board's careful attention. Like the technology itself, insurers' understanding of the impact of digitisation is evolving rapidly and is destined to deeply modify the whole financial and insurance ecosystem, impacting all points along the insurance value chain, from underwriting and risk management to distribution and claims, and consequently reshaping the competitive landscape and customer relationship. Information Technology devices are essential not only in the productive and distributive processes, but also for a more efficient and effective setting of the decision-making process and organisational structure.

Many technological innovations can affect insurance business models, which can be used both in the back end and in the front end. We are here referring to the blockchain technology, to the artificial intelligence, to robo-advisors and other systems of record for the core insurance business and its support. Digitisation helps insurance companies in designing new products and in calculating the prices of new and existing products. Digital technology changes the quality and the ways data are analysed, on the basis of which the risk assessment is carried out, thus allowing insurers to better profile the customer in order to determine the risk pool.

Within the scope of distribution, however, many new technologies are evolving, offering new options for consumer interaction. Technological innovations, which are applied in the insurance delivery process, permit insurers to interface with customers and to offer a rich multichannel, multidevice digital customer experience.

Digital technology, breaking down entrance barriers, is encouraging new entrants into the insurance sector. Notably, in recent years there has been a sharp pick-up of high-tech start-up firms—or InsurTech—particularly in personal lines and distribution. The InsurTech development should represent a threat for traditional companies, from Big Data analysis to digital devices, from personal interactivity to home automation systems development. However, a drastic disintermediation of insurance companies, which would imply a strong innovation of business models from incumbents, does not seem to show up in the short–medium term. The majority of digital competitors, at least at the moment, is not in a disruptive position, and is rather in an enabler position: thanks to their technological competences, they can facilitate and improve the efficiency of the traditional insurance business. Incumbent insurers and InsurTech start-ups have much to gain from their collaboration. InsurTech start-ups and incumbent insurers have complementary strengths, with InsurTech offering better value for money and timely and efficient service, and insurers offering superior security, brand identity, and support for personal interaction.

The trend to digitisation is most notable on the services side of distribution. The new digitised technologies and the new habits of customers bring about major changes in the ways insurance services are offered and used by customers. Digital platforms and systems create direct channels with the user, increasingly reducing the need for intermediaries like agents and brokers.

If, on the one side, the new technologies reduce the personal contact with the customer, on the other side they permit, especially for less complex products, to increase the frequency of contacts with the latter, thus offering the possibility to increase the customer's loyalty.

Insurance companies are accelerating the shift to a radically different delivery model to make their services more available to customers, attributing an increasingly important role to technology in the majority of interactions. This will have implications for insurers' business models, in the way they interact with their customers and the nature of the products and services they provide.

Surveys highlight that in many countries the traditional intermediaries still represent the main distribution channel of insurance services. This happens especially for services with a higher added value and a higher level of complexity, for which the personal contact or the advice of agents or brokers are essential. In these areas, technology is being applied to improve the efficiency and effectiveness of agents and brokers. However, many consumers want a seamless shopping experience anytime, anywhere. They are more self-directed in their insurance decisions and want to interact with their agent across a full range of channels: in person, through mobile devices, by phone, Internet, and video conferencing, when researching and buying insurance. To this end, the development of robo-advisors, which use artificial intelligence to formulate automated advice and recommendations, could facilitate a further e-commerce penetration into insurance, and also reduce operational costs.

Traditional players will also have to reach for the digital interaction, if they want to maintain a relationship with more evolved customers, without overlooking, in this respect, the expectations of the so-called millennials, who represent future customers. These consumers represent the first generation in history that are strongly familiar with digital technology and spontaneously know its communication codes. Their experiences with the ever-present online and app-based consumer environment influence their expectations for purchasing insurance. This is a fundamental change in the nature of insurer–customer interactions.

In this respect, it is necessary to point out that the increasing automation of innovative distributive channels, if not used in an interactive way, contributes to a progressive depersonalisation of the insurer/consumer relationship. It is also important to highlight that certain circumstances, such as the increasing competition among insurance institutions, as well as the economic-cultural evolution of the customers, contribute to increase the mobility of the latter, thus also reducing the intensity of the insurance relationship. A clear example is represented by price comparison websites, which provide consumers with more information on products and costs, often selling a product directly, with no agent or broker involvement. This increases choice opportunities and the likelihood of switching from one offer to another, thus reducing the customer's loyalty, as convenience judgments are increasingly expressed on the basis of technical-economic reasons, and less on the basis of emotional factors.

The transformation of consumption patterns towards the digital economy leads to the need to modify the traditional ways insurance services are provided, as well as the consumer engagement model.

Insurers who fail to meaningfully differentiate their offerings will suffer from a lack of consumer–buyer engagement, give up business to competitors, and leave themselves vulnerable to disruptive entrants. Slow-to-adapt incumbents who insist on viewing their products as commodities, competing only on price, will not be able to succeed against those able to adopt a buyer-driven approach, learning how to attract and retain customers through brand differentiation and customer-centric capabilities.

It is therefore necessary to seek a continuous improvement in the communicational approach with the reference market in order to try to renew and innovate, where it is possible, the customer relationship to increase customer loyalty and retention.

The book is divided into six chapters including the Introduction. Chapter 2 aims to analyse, on a prospective basis, the most relevant issues relating to the changes, the opportunities, and the challenges posed to insurance managements by the use of new technologies, with a particular focus on the most significant aspects of the distribution of insurance products also from the regulatory point of view, in the light of the imminent entry into force of *Insurance Distribution Directive* (Directive 2016/97/EU) which, in an innovative way, sets up *Product Oversight and Governance arrangements* and product governance obligations since the moment of the product design.

Chapter 3 highlights the disruptive phenomenon of the diffusion of the InsurTech start-ups, trying to focus on the genesis and the peculiar characteristics; it ends focusing on which should be the strategic response levers of the incumbent carriers to the digital transformation perspective of the entire business model.

Chapter 4, after analysing the peculiar characteristics of the delivery system of insurance services, focuses on the study of the distribution channels of a technological nature that have a greater degree of innovation in the insurance sector, deepening risks and opportunities.

Chapter 5, in the light of the depersonalisation of the customer relationship due to the spread of the technology, focuses on the importance of direct marketing and effective social media strategy to improve the relationship between company, channels, and customers.

Chapter 6 reports the results from a survey conducted on a sample of insurance companies located in Europe and USA to test the main features of their websites distribution, highlighting their distinctive traits under the three main aspects of accessibility, transparency, and quality of the online distribution.

CHAPTER 2

Technology and Insurance

Abstract The chapter highlights how digitalisation is destined to deeply modify the whole financial and insurance ecosystem, impacting all points along the insurance value chain and consequently re-shaping the competitive landscape. It focuses on how the increasing collection of data through the new digital technologies permits more granular underwritings in risk insurances. Smart analysis techniques, predictive modelling and connected telematics devices allow insurers to create products and set premiums based on actual risk profiles rather than on general standards. The delivery process is also affected by the new technologies. This is also taken into account by the regulatory framework of the Directive 2016/97/EU, *Insurance Distribution Directive*, which is soon to be introduced at a European level.

Keywords Insurance value chain • Insurance sales and distribution • Technology and insurance distribution channels • Insurance distribution regulation

1 INTRODUCTION

Technological trends and changes in consumer behaviour are encouraging companies to consider new business models. Digitalisation is destined to deeply modify the whole financial and insurance ecosystem, impacting all points along the insurance value chain and consequently reshaping the

© The Author(s) 2018

A. Cappiello, *Technology and the Insurance Industry*,

https://doi.org/10.1007/978-3-319-74712-5_2

competitive landscape. The increasing collection of data through the new digital technologies permits more granular underwritings in risk insurances. Smart analysis techniques, predictive modelling and connected telematics devices allow insurers to create products and set premiums based on actual risk profiles rather than on general standards.

The reduction in profit margins, due to the increased regulation and growing competition from new entrants on the market and incumbent companies, pushes the latter to seek, on the one hand, cost savings and improved efficiencies, and, on the other hand, a better competitive effectiveness through strategies of greater proximity to customers and of customers' loyalty.

New technologies deeply affect the delivery process, in which the use of digital solutions and the emergence of new channels deeply change the sales process of the service, the mode of use, and consequently the relationship with the customer. About the latter, the new digitised terrain will also present challenges for many insurers. The evolution of the digital economy is radically changing customer expectations and behaviour. Greater use of portals and aggregators by customers, for example, will increase product transparency and reduce costs of switching insurers. Consumers have a wide range of information available to evaluate their risk exposures and are becoming more self-directed in how they choose to fulfil their insurance needs. Consequently, the way companies conduct their business to engage their customers needs to change. Customers demand more individualised offering, fully integrated sales and service channels.

2   THE IMPACT OF INFORMATION TECHNOLOGY ON INSURANCE

The evolution of the competitiveness of financial markets highlights the strategic role of the most advanced computer technology: these are the contributing cause of changes in competitive scenarios as well as the instrumental factor in the achievement of the business objectives of the new environmental context.

As margins decline and price competition increases, insurers must compensate by improving operational efficiency and controlling the costs. They require to balance the need for innovation and protection against risk, remain compliant with government regulations, and assure security across all functional areas.

To address all these challenges, insurers are moving to digital-ready infrastructures that enable a new level of real-time process and interaction enabled within new business models. A well-considered technology investment plan based on personalisation, workforce innovation, optimised business operations, and managing risk capabilities allows to generate savings and grow revenue over the course of the digital transformation journey.

Today's insurance companies face a time of tremendous change, thanks to shifting economic circumstances, changing customer expectations, and an increased competition from new players in the market. Companies must meet the demands of increasing regulation, new types of security threats, game-changing technology, and an ever-younger workforce. To meet these challenges, the insurance industry is turning to digital transformation, adapting innovative business models similar to those increasingly seen in other industries (Morgan Stanley 2015; Cavina et al. 2017).

Technology, on the one hand, allows the reduction of production and distribution costs (thus improving operational efficiency) and, on the other hand, modifies the products themselves until reaching innovation of the same and their systems of delivery, with notable advantages in competitive terms; all this needs to be put in relation to the progress of the public's economic culture and to the changing needs of the consumer (Cappiello 2012).

Technology, in addition to widening volumes and geographic areas, also allows us to use more flexible structures and offers services that are more responsive to consumer expectations, even regarding negotiated prices (Manning et al. 2014).

It follows that technological innovation revolutionises the entire business to a large extent, as it plays an incisive role in every area of business; the new possibilities for management and data processing also contribute to the rationalisation of the organisational structure and the decision system.

Therefore, it can be assumed that the use of advanced technologies involves and modifies the following aspects of insurance management: (1) productive and/or distributive process, (2) organisational structure, and (3) decision-making system.

Obviously, due to on going changes, in an increasingly complex environment, information technology (IT) tools become indispensable not only in the field of productive and distributive processes, but also for the purpose of a more efficient and effective decision-making process and organisational arrangements (IIF 2016).

We can refer, for example, to the introduction of the decision support systems into insurance management, as well as of the most advanced artificial intelligence systems (Lamberdon et al. 2016; McKinsey 2016a).

Obviously, potential applications are countless, and all of them are undoubtedly of interest, as these tools contribute, among other things, to the emergence of new professional roles of strategic importance.

At a strategic level, IT facilitates the rationalisation of decision-making on the activities of planning and control, on human resources management and planning of various marketing actions.

For evidence, the standardisation of decision-making processes, and thus the homogeneity of the behaviours made possible by the mentioned applications, permit the adoption of operational and decision-making solutions, having a strong impact on the organisational structure of the company activities.

It should be noted, moreover, how the adoption of IT structures capable of interacting with special "support" centres, where the production phases not requiring the presence of the customer take place in a centralised manner, permits to obtain efficiency gains through the rationalisation of operating procedures (McKinsey 2016b).

If, therefore, it can be said that IT is an undoubtedly effective strategic factor, as well as an element of achieving economies of scale and increasing productivity, it's important to point out that IT must find a placement of integrated type at all levels of company functions. Otherwise, investments in technology, poorly coordinated at the system level, wouldn't be able to cope with the competitive challenges that the insurance industry will face in the years to come.

In this respect, the close relationship between technological policies, business strategies, and organisational structures shouldn't be neglected.

Conscious management of change requires therefore a close correlation between the overall corporate strategy and use of IT strategies in different areas of the insurance business, to increase efficiency, improve consumer relationship, and consolidate or increase the market share (Scardovi 2017; Willis Tower Watson 2017).

Global insurers are unparalleled in their ability to assess and manage risk, having fine-tuned their underwriting expertise over decades. Currently, the dissemination of new technologies is forcing them to alter long-held business practices. Insurers have to deal with a flood of new technologies, with range from wearable devices to driverless cars, and are expected to be taken up by a large number of consumers in coming years. With the fast advancement of such technologies, insurers must develop strategies that take full advantage of the opportunities they present, while minimising the risks (Schmidt et al. 2017).

The range of technological innovations that will be likely to affect the insurance business models is wide. For example, in the field of property and health insurance, wearable devices capable of tracing various parameters—from vital parameters to the sleep cycle—as well as the ability to monitor driving style at a distance, allow us to collect a wide range of data related to risk assessment and premium calculations. Similarly, advanced medical technology is making healthcare more proactive and reliable, changing the metrics by which insurers assess health. As the data generated by these advanced technologies become connected via the Internet of Things (IoT), the amount of insight that can be developed from the data grows powerful (Haddud et al. 2017). On the one hand, insurers can use increasingly advanced tools to quickly analyse volumes of data coming from various sources and drive actionable insight in real time. On the other hand, benefits include the use of robotic process automation, which is ideal for handling insurers' many rules-based administrative tasks (Craneld and White 2016; Deloitte 2015; Keller and Hott 2015).

Many of the innovative technologies can be used both in the back end and in the front end for process optimisation through the value chain (Porter 1985, 1998).

The first case concerns the use of blockchain, artificial intelligence, advanced analytics, robotic process automation, and other systems of record for the core insurance business (the policy administration, claims, and billing functions) and its support (e.g. risk management and finance). Decision engines and artificial intelligence support decision-making, allowing insurers to propose tailored customer-centric services based on micro-segments and personalised risk profiles. In contrast, legacy systems, in which core business processes (such as pricing and underwriting) are "hard coded," allow for only static decision-making based on broad customer segments and statistical patterns. Digital platforms integrate modular product architectures and "zero touch" processes. The former enables insurers to package multiple product and service components into a broad customer proposition, while the latter are completely automated processes that can be changed with minimal involvement from IT.

On the other hand, another category of technological innovation is applied in the distribution process of the insurance product: we refer to the use of devices, context-aware and location aware services that allow insurers to deal with the customer and offer advice tailored to the needs of the latter, as well as a rich multichannel, multidevice digital customer experience.

3   Technology and Insurance Value Chain

For a long time, the traditional insurance business model has proved to be remarkably resilient, but the time has come for the insurance industry to face a profound technological change. While some aspects of technological change—such as better operating efficiency, the need to engage with consumers digitally, and increased disintermediation—are common to many industries, others are specific to insurance.

Insurers have always been intense users of data in analysing and measuring the risks they underwrite, setting the associated terms and conditions for insurance policies, assessing risk, and claims management.

Digital technology changes the entire insurance value chain; it changes the type of data that insurers use to assess risk, the way in which information is analysed, and, ultimately, the size of the actual risk pools (Table 2.1).

The notion that insurance is a low-engagement, disintermediated category in which customer relationships can be delegated to agents and intermediaries is now outdated. Instead, digital technology and the data and analysis it makes available give insurers the chance to know their customers better, which means that they can price and underwrite more accurately, and better identify fraudulent claims (Guha et al. 2015). They can also offer clients more tailored products and in a more timely manner.

From what is said, it turns out that Big Data, artificial intelligence/cognitive computing, predictive modelling, wearable devices, telematics, and the IoT are having impacts all along the insurance value chain, enabling new

Table 2.1   Impact of digitalisation on the insurance value chain

Product development	• The use of Big Data facilitates new behavioural, granular data collection and enables service personalisation • Telematics may reduce associated risks but create new ones, such as cyber risk
Sales and distribution	• Comparison platforms present customers with a comprehensive choice of all kinds of insurance covers and in some cases allow to buy insurance online • InsurTech start-ups entry in the insurance market from adjacent markets
Underwriting	• Instantaneous information and Big Data allow more predictive and evaluative analytics • Finer segmentation is driven by greater processing capabilities
Claims	• Telematics provides instantaneous information which can help insurers with more accurate claims assessment and reduce fraud • Technology decreases processing time

ways of communicating, information sharing, and insuring (McKinsey 2010). All aspects of the insurance value chain are affected by information—from administration to pricing/underwriting and distribution (Rayport and Sviokla 1995; Meier and Stormer 2009).

IoT and connected sensors could revolutionise the product design opening up many new opportunities in connected home and connected health solutions. Cheap, connected monitoring devices offer a fundamentally different way of assessing and, crucially, mitigating risk (Porter and Heppelmann 2014; IHS Markit 2016).

In such a scenario, new risks emerge to replace old ones. Ever since insurance was established as an industry, the types of risk that individuals and businesses face have been changing. Over time, new risks have always raced to replace old ones. Some risks we can't conceive today may be important for insurers in the future (Venture Scanner 2016; IAIS 2017).

Technology has the capacity to reduce existing risk pools or to expand them—with strong implications for insurers. The smaller the risk pool, the smaller the premiums it can generate, affecting the revenues of the sector.

The global consumer survey of Morgan Stanley (2014) finds that the motor risk pool could shrink by about 5–9% of total global non-life premiums (excluding health) over the next 10 years. This is because technology reduces the rate of fraud and reduces the accident rate, thanks to driver monitoring, feedback, and education. On the other hand, other risks increase or occur for the first time, such as cyber risk, mobile phone radiation risks, and the risk of contact with nanotechnology materials. This must be carefully evaluated, particularly at a time when low interest rates and tighter regulation constrain performance (Ernest Young 2015; AXA 2017).

New technologies make policy and claim management more efficient, as machine learning and pattern recognition are used to analyse handwritten and unstructured documents to expedite and detect false claims. Insurance claims can be processed via online platforms, with less time for processing (Hook 2016; OECD 2017). Automation, analytics, and consumer preferences are transforming claims processes, enabling insurers to improve fraud detection, cut loss-adjustment costs, and eliminate many human interactions.

Insurers are also experimenting with blockchain technology—digital distributed ledgers which are cryptographically safe—to improve the efficiency of processes within and among existing institutions, such as in claim management or reinsurance contracts. Blockchains offer benefits of speedier connectivity between counterparties and potential for reduced fraud or loss-adjustment expenses, all of which help lower insurers' overall costs (McKinsey 2016c).

In addition to product design and claim management, the digitalisation is helping in the pricing of new and existing insurance products.

The insurance market is characterised by information asymmetries. From the point of view of insurers, these mainly concern the need to find adequate information on the costumer's risk profile. A decisive factor in the success of a business model in insurance is the insurer's ability to estimate the cost of risks as accurately as possible. While in the case of some simple product lines, such as motor insurance, the risk cost estimate may be widely or fully automated and managed internally, in areas with complex risks, the help of a third-party expert can attenuate this type of information asymmetry.

The growing proliferation of new data about insureds collected via sensors and smart devices permits more granular underwriting of individual risks. Smart analytics is the term used to describe the systematic use of data. To this end, both structured data from a range of sources (such as sensors, written documents, and data published on the Internet) and unstructured data gleaned from conversations or letters are gathered. Software solutions recognise patterns in the data and cluster them, giving insurance companies more detailed insights into the behaviour and needs of their customers.

Smart analytics, predictive modelling, and connected telematics devices allow insurers to create products and set premiums based on how insureds actually behave rather than using general proxies. As new hazards are identified in real time, insurers can improve their data sets to better manage eligibility, underwriting, and rating (PwC 2014).

Normally, the insurance industry relies on historical data to assess client risks. Once underwritten, the client pays an agreed premium throughout the policy period irrespective of risk variations (Paci 2017). Thanks to the use of telematics and wearable technologies, now, during the policy period, it's possible to gather in real time some parameters as the driving behaviours and health metrics of the insured. The combination of rich customer data, telematics, and enhanced computing power permits to adopt premiums based on the actual use and on the conduct of the insured. Insurers can use the real-time data captured through telematics devices and powerful analytics to reassess the current risks and recalculate the premium for current risks at regular intervals. Insurers can develop tailor-made products with pricing adjusted to individual risk levels and very accurate selection (Porter and Heppelmann 2015).

Risk selection is becoming ever more accurate and precise. The ways in which technology is improving, or may in the future improve, risk selection include (1) the use of data gathered from connected sensors (IoT); (2) the use of Big Data to enriching the underwriting decision; (3) forward-looking, sophisticated measurement of risk (cat modelling); and (4) digitalisation of insurance, which makes data more readily analysed and products more readily adapted (Morgan Stanley 2014).

The cost of improving risk selection has to be weighed against the benefits. The question is whether the additional advantages in lower loss ratio can outweigh the costs of investing in new technology that allows better risk selection.

Increased digital interaction with consumers has facilitated the capture, storage, and management of large quantities of data about customers. Using analytical techniques to extract business intelligence from this information is often collectively called Big Data, although the term is not always used consistently.

For insurers, it offers the opportunity to assess their customers' needs, target products and services to individuals and businesses, support underwriting decisions, and reduce the cost of fraudulent insurance claims (Laskowski 2013). At the same time, it entails risks relating to the permissible and appropriate use and management of customers' data as well as the challenge of designing business processes and products that will provide a profitable return on investment.

The importance of Big Data lies not just with the collection and storage of large and disparate pieces of information, but also in the ability to analyse and extract tangible and useful knowledge from that data. Also, some researchers believe that by combining more and different types of information, you can reduce problems related to incompleteness of particular pieces of data (including the likelihood of errors, inconsistency of formats, inaccuracy of data processing, etc.) (Cukier and Mayer-Schenberger 2013).

Some insurers have gone as far as to introduce fully automated underwriting systems which provide final decisions on life insurance applications without intervention by a live underwriter. Big Data can additionally help businesses to improve other core functions, including marketing, distribution, operations, and claims. Real-time predictive analytics offer insurers the chance to respond rapidly to changing customer behaviour.

3.1   Focus on Changes in Insurance Sales and Distribution

A great impact on technology is on the distribution of insurance products, where new bid methodologies and new channels deeply change the service delivery process, the way they are used and, consequently, the customer relationship (Capgemini/Efma 2016).

Internet is changing almost all the stages of the distribution process, primarily collecting preliminary information from the consumer. Other pre-sale activities such as consulting and trading tend to move to the web, albeit to a lesser extent (Swiss Re 2014). The completion of the sales process is also possible online for some types of policies.

The use of technological solutions reduces the cost of matching demand and supply and information asymmetries typical of insurance intermediation.

Among the costs borne by the supply, we refer to the management and administration costs associated with the distribution of traditional products. These include (1) the costs of setting up a branch; (2) the training costs of sales employee; (3) fees paid to intermediaries; and (4) administrative costs and others.

Evidently, direct sales solutions are examples of distribution models that can cut the costs mentioned earlier and reach a much wider audience. This latter aspect is crucial in the field of insurance management, where the principle of solidarity that underpins the concept of insurance can only work in the presence of high operating volumes.

Using many new technologies, on the one hand, reduces personal contact with the customer and is, on the other hand, able to offer new opportunities to increase the frequency of consumer interaction, especially for the benefit of less complex services, with obvious advantages also in terms of loyalty.

Surveys indicate that consumers often continue to value the personal interaction and expert advice of agents and brokers, especially when it comes to complex insurance for commercial, financial, and life and health risks. In many countries, traditional intermediaries still represent the dominant channel through which insurance policies are sold: in these areas, technology is being applied to improve the efficiency and effectiveness of agents and brokers.

However, many individuals want a seamless shopping experience anytime, anywhere, whether online, by phone or in a store or agent's office. To this end, the development of robo-advisors, which use artificial intelligence to formulate automated advice and recommendations, could facilitate further e-commerce penetration in insurance and, also, reduce operational costs.

Consequently, the use of digital technologies can be positive, in order to reduce the loyalty costs of less sophisticated services, but also to finalise cross-selling policies by offering more and more personalised products according to the needs of the insured.

Because of the information asymmetries that characterise insurance intermediation, insurers may have limited access to customer information about their tastes, financial needs, consumption, and buying behaviour. Information asymmetries may be greater in case of sales through agents or brokers where client ownership is not entirely in the hands of insurers (Colombini 2008).

Internet, mobile devices, and telematics are radically changing the relationship between insurers and customers as a policy can be sold directly without the involvement of an intermediary. Regarding this, the use of Big Data facilitates a much deeper understanding of customer wants, needs, and behaviours. Insurers have more opportunities to observe customer interactions at different points in the distribution process rather than rely on knowledge and insights from agents, intermediaries, and company employees. This means they can gain a more holistic view of consumer preferences and behaviours and use this information to become more consumer-centric in their distribution activities (Swiss Re 2014).

To define a more granular classification of existing and prospective customers, using Big Data allows you to extrapolate many new information for the customer segmentation. These data concern activity-based data, such as website tracking information, vehicle telematics about customer driving behaviour, purchase histories, call centre, and mobile data; social network profiles (e.g. work history and group membership); and social influence and sentiment data, such as product and company associations, online comments and reviews, and customer service records.

This micro-segmentation enables ever finer targeting of content, offers, products, and services, which can deliver real and substantial returns.

For example, insurers can also use the web and social profiling to identify which sections of the community are seeking but failing to find insurance cover. Or, they can combine information on customers' retail shopping habits with claims histories to target prospective customers, or, finally, employ telematics to improve auto underwriting and customer engagement. Insurers are increasingly looking to extract value from telematics both in terms of assessing underwriting risks and the potential opportunities to cross-sell and up-sell.

The use of advanced statistical techniques and data analysis to evaluate the impact of multiple explanatory factors on a particular variable can be used to target customers, personalise insurance products and services, and anticipate customer needs and their likely actions. This can have a better sell rate than broad-based campaigns such as advertising. It can help insurers to cross-up/sell additional product and policy features and reduce marketing costs.

On the other hand, between the costs incurred by the demand in the buying process and the insurance service use, we detect the costs of research and processing of information relating to various insurance options and the detailed information asymmetries.

Having access to the entire market and being aware of a wide range of available products is a prerequisite for consumers in order to make good purchasing decisions. Consumers may not be sufficiently sure about buying choices until they think they have explored a sufficiently wide range of options.

Potential transactions may not simply exist due to informational asymmetries that don't allow the consumer to have adequate knowledge of the entire range of available products, and the suitability of certain products to meet certain risk coverage requirements.

Such asymmetries may be more accentuated in the presence of complex long-lasting insurance products (e.g. life insurance), where the suitability of a product can be assessed only with difficulty and, potentially, only at point in time after the subscription of the policy; even after the policy expires or when the accident occurs (Shamdami et al. 2008).

Even when access to the entire market is possible, the ability to make an effective comparison between available options requires time and costs depending on the nature of the market. In case of relatively standardised insurance products, where purchasing decisions are primarily attributable to the price, processing costs can be quite limited. However, these costs are higher in the case of relatively more complex products (Rawson et al. 2013; McKinsey 2015).

There are different ways in which consumers balance quality and price, and comparisons between differentiated products (i.e. products that offer a different price/quality mix) can be more problematic. In such cases, personal interaction with professional intermediaries providing expert advice can add considerable value to the consumer buying experience; agents and brokers, especially in the context of commercial lines, can provide a valuable contribution.

For products that are less complex, consumers expect personalised, self-directed interactions with companies via any device at any hour, as much as they do with online retail leaders. Distribution channels are responding to changes in consumer preferences. Policyholders increasingly demand digital-first distribution models in personal and small commercial lines, while aggregators continue to pilot direct-to-consumer insurance sales.

Price comparison websites, which have been around for quite some time, are providing consumers with more information on products and costs, especially for more commoditised products like auto and travel insurance. They often sell a product directly with no agent or intermediary involvement. Modern consumers are more self-directed in their insurance decisions and want to interact through various channels when researching and buying insurance.

Consider that the likelihood of operations is greater if consumers think they are receiving the service from trusted suppliers selling products under transparent contractual terms. In this way, cross-buying opportunities are also created: where such confidence is an important factor, cross-buying (i.e. the purchase of various products and financial services from a single point of sale) is more likely.

Likewise, the rise of the Internet as a distribution channel depends on the growing consumer confidence in online security, varying from country to country, as well as from person to person.

Finally, information asymmetries also affect the nature of the sale and the quality of the advice received. For example, consumers may not know the existing contractual relationship between an intermediary and insurers (whether or not they're authorised representatives) and not be able to properly assess the impact that the remuneration methods of the intermediary (e.g. of commission based on sales volumes) may have on the quality and/or impartiality of the advice required.

The way in which the insurance distribution is changing offers great future opportunities for insurance companies, but it also requires insurance companies to deal with major issues.

We are here referring, among other things, to the economic benefits of implementing new technologies; to addressing the increasingly stringent regulatory cost of insurance distribution; to the management of new governance relationships between insurers and intermediaries; and to the proper management of Big Data to track, respond to, and, if possible, anticipate consumer expectations and behaviours (Bughin et al. 2017).

However, insurers will need to be careful to not alienate customers by taking product personalisation too far. There is in fact a fine line between Big Data marketing applications that provide relevant product and service recommendations and those that appear intrusive. Targeted marketing may be perceived as invading individuals' privacy, especially when purchase recommendations are generated and sent without a consumer's conscious opt in. Insurers must understand the limits to data-driven personalisation, or else they might alienate potential customers and in a worst case face legal risks (Swiss Re 2014).

4   THE REGULATION OF INSURANCE DISTRIBUTION: THE DIRECTIVE 2016/97/EU

In recent years, the insurance industry has been subjected to radical regulatory and supervisory changes that have created significant operational challenges.

First, for European companies we refer to the introduction of supervisory regulation enshrined in the risk-sensitive framework of Solvency II, which required insurers to make substantial investments in improving systems and processes.

Furthermore, in several markets we see a trend towards enhanced consumer protection (ASIC 2016). Regulators are concerned with several aspects of consumer protection: bans on commissions and increased requirements on advisory services; data privacy and protection legislation; and anti-discrimination requirements (Lannoo 2017; Pradier and Chneiweiss 2017).

Recent EU legislation—such as the Insurance Distribution Directive (IDD) and the Packaged Retail and Insurance-based Investment Products (PRIIPs) Regulation—is likely to have an impact on existing distribution structures, as are regulatory developments in individual EU countries (Biener et al. 2013; Douady et al. 2017).

PRIIPs is an European regulation that gives consumers more protection and allows for like-to-like comparison of retail investment products, making them more transparent. In the light of the new regulation, insurance, asset management, as well as investment banking companies will be required to provide investors with a PRIIPs key information document prior to investment.

On the other hand, the European Council amended the legislation on insurance distribution by approving the Directive 2016/97/EU (IDD), recasting the former Directive 2002/92/EC on the distribution of insurance products which is lacking, over time, in practice.

The new Directive, which is soon to be introduced, contains important rules that aim to standardise the sector by creating common standards for the sale and disclosure of information to customers, and to strengthen consumer protection in this respect.

In general terms, it should be noted that the IDD is based entirely on the notion of insurance *distribution*, instead of the *intermediation* one used by the previous Directive.

This is not a merely lexical nuance, but the means to expand the circle of recipients of the provisions, which are therefore also addressed to all agents considered *ancillary* (and therefore insurance intermediaries in the strict sense) that, for various reasons, participate in the sale of insurance products, including direct sales networks of companies and price comparison websites that can distribute policies directly or indirectly. With reference to the latter, it should be noted that in the 2013 European Insurance and Occupational Pensions Authority (EIOPA) had taken note of their importance in the market, as well as the lack of uniform rules and the various demands aimed at bringing those persons among the recipients of the provisions on insurance distribution.

The purpose of the subjective application extension is to be found, on the one hand, in the declared need to ensure uniformity of consumer protection regardless of the proposer of the insurance product, and, on the other hand, in the need to standardise the treatment of different subjects involved, to varying capacity, in insurance distribution, avoiding distorting effects on competition (Box 2.1).

However, it should be noted that the IDD is a minimum harmonisation directive, while retaining the power of Member States to maintain or adopt more stringent consumer protection provisions when justified by the national context.

Box 2.1   Main Objectives of the Insurance Distribution Directive

- Improve *retail insurance* regulation, so as to facilitate the integration of the single European market.
- Establish the conditions for a fair competition between insurance product distributors.
- Increase the degree of consumer protection, especially regarding life insurance products investment.
- Extend the scope of the rules to all distribution channels, applying them proportionally also to those who make the sale of insurance products on an ancillary basis.
- Identify, manage, and mitigate conflicts of interest.
- Tighten up administrative sanctions and the measures to be applied in the event of non-compliance with the regulatory requirements.
- Improve the adequacy and objectivity of insurance advisory services.
- Effectively combine the professionalism of salespeople and the level of complexity of insurance products.
- Ensure greater clarity regarding market access procedures in the case of cross-border insurance products.

A topic that deserves study is certainly the one related to the provisions on organisational requirements for governance and control of the product (*product oversight and governance arrangements* [POG]) provided for by Article 25 of the Directive, which introduces the *product governance* obligations for producers and distributors who develop and market insurance products.

These provisions require the existence of consumer protection tools from the time of design and for the entire duration of the product life cycle, providing for constant monitoring to ensure that it continues to meet the purchase objectives of the intended target market.

Specifically, the Directive requires that the company and the intermediaries who, together with the latter, make an insurance product, adopt and manage a formal process, aimed at the preventive approval and definition of the relevant content and any eventual, significant change before the distribution stage. These procedures require a careful analysis of the product characteristics, in relation to a specific customer target and to a coherent distribution strategy (agency or ancillary), and the adoption of testing measures and periodic review criteria both of the products and of the same procedure, to ensure the ongoing consistency of the strategic management of the product/channel/customer relationship.

The adoption of the dispositions set out under the Directive on product *governance* will ensure that the adequacy of the offer to the user's characteristics is not addressed exclusively in the distribution phase, with verification responsibilities borne by the intermediary, but becomes a precise *product development* business of the insurance company.

The reference to the *target market* provides that the product manufacturer develops and provides insurance instruments with characteristics tailored to the needs and goals of the customer segment identified as a target. The product manufacturer will have to consider, in the product study, the level of knowledge and financial education of the target audience, with the responsibility to proactively identify groups of users for whom the product is inappropriate in terms of potential targets or financial characteristics. The product/policy adequacy evaluation is therefore reinforced in the upstream phase of the distribution, with direct responsibility on the manufacturer to the product formulation level, and responsibility on the part of the distributors control at the sale stage.

It has to be noted that, in view of the new provisions on POG, for certain types of life policies of saving investment characterised by apparent financial complexity, it would be quite difficult to extend the *target market* to all retail customers, as it actually happens today, with the consequence that, in a forthcoming IDD operation scenario, the company will be required to offer them only to a predefined customer segment.

In order to ensure proper compliance with the expectations/needs of the customer, the Directive provides a constant monitoring by intermediaries not only with respect to the products at an early designing stage, but also for those products which are already being distributed and for those undergoing restyling. Regarding this, the distributor plays a key role, as he represents a direct contact with the market, and can also be an active participant in the design, development, and redesign of an insurance product for the reference target.

According to EU rules, intermediaries/distributors will have to define a steady flow of information with the company in order to report cases of inadequacy of the products. Since the information about the customer constitute an element of increasing importance, in view of an increased protection of insureds, it will be necessary to implement effective *reporting* systems, in order to guarantee the validity and timeliness of the adequacy verification processes by the relevant offices, such as the actuarial office.

Within the scope of the IDD, the general information obligations are essentially unchanged from the current regulations. Instead, the reference to pre-contractual information requirements constitutes a novelty item, affecting various aspects of the distribution process to enable the customer to make informed decisions.

First, the Directive sets disclosure requirements on the status of those responsible for the provision of insurance products and, in particular, Article 19 sets out—with no distinction between products of life insurance and non-life insurance products, and, with respect to the latter, without distinction based on the nature of the risk—the intermediary's obligation to provide information about the nature of its fees, if fixed on an hourly basis, if of provisional nature and/or otherwise. The Directive, for the first time, introduces the obligation to structure remuneration policies that avoid negative impacts on the quality of the service.

Increased consumer protection requires "to place" the product only in compliance with the insurance requirements of the customer (*demands and needs test*), so as to prevent or mitigate any phenomena of *mis-selling*. In this light, Article 20 introduces two distinct documents, and in particular: (a) the *personalised recommendation*, required when consulting activities are carried out, and specifically meant to identify the reasons why a particular product may be considered more appropriate than others to meet customers' needs, and (b) for non-life products, a standardised, easy-to-read information document containing basic information.

In particular, the Directive lays down certain specific obligations for the distribution of *insurance-based investment products* (IBIPs) that have a maturity or redemption value exposed, in whole or in part, to market fluctuations.

The provisions relating to these products provide obligations regarding conflicts of interest, the information on products and associated costs and charges, and the assessment of the product in case of advisory service.

First, insurance companies and insurance intermediaries are obliged to adopt and maintain effective procedures for identifying and managing conflicts of interest in order to prevent them from harming their customers. If the procedures are not sufficient to prevent injury to the latter, companies or insurance intermediaries are required to inform the customer of that fact before the contract is concluded.

As for the additional disclosure requirements, insurance companies and intermediaries must provide appropriate customer information at least about (1) the periodic assessment of the suitability of the products in case of distribution associated with the advice; (2) guidelines and warnings on the risks associated with products or investment strategies proposed; (3) costs and charges, including advice costs (if provided); and (4) payment methods.

Article 30 is also in charge of the suitability assessment requirements and adequacy of IBIPs in the event of the consulting service, defined as providing customised recommendations to a customer, on the customer's request or on the initiative of the insurance products distributor, with respect to one or more insurance contracts.

Notwithstanding the general informative obligations in Article 20, insurance intermediaries and investment companies that carry out consulting activities must (1) obtain from customers information on knowledge and experience in the relevant investment field for the specific type of product or service, the financial situation and the investment objectives; and (2) recommend appropriate products with respect to the risk predisposition and the customer's ability to sustain losses.

When, on the other hand, no consulting activity is provided, intermediaries and companies limit their request for information to the sole knowledge and experience of the client.

Similarly to the dispositions of the previous directive on the placement of insurance services, the distributor will have to alert the customer if, on the basis of the information gathered, he evaluates the product as inadequate or inappropriate to his investment requirements and objectives. If the customer doesn't provide the requested information, he will be warned of the inability to receive information on the appropriateness or suitability of the product.

The Directive also provides for the possibility for Member States to introduce a simplified distribution system for intermediaries or non-consulting companies, a reminder of the *only execution* service of the financial intermediation. In particular, intermediaries and companies may be exempt from the obligation of collecting information on customer knowledge and experience in cases where (1) products have underlying non-complex financial instruments, (2) distribution is made at customer's initiative, (3) the customer has been informed that he/she has lower guarantees as the intermediary is not obliged to assess the adequacy of the product, (4) the intermediary or the company has fulfilled his obligations regarding the management of conflicts of interest.

As we have just briefly pointed out, the approach of the European legislator is clear in redesigning relations between customers, agents, and companies, as expected regarding the skills and knowledge required for the entire distribution chain. It's, however, more difficult to understand which path the various parties will have to undertake to achieve the goal proposed by the IDD, since the regulatory framework is not fully defined, and the impact and magnitude of these innovations will also depend on the content of the second-level regulations, made up of delegated acts, technical advice, and guidelines as well as the implementation that will be given at national level (EIOPA 2017).

In the current scenario, the challenge is the need not only of a regulatory kind, but also, above all, of the market, of a real evolution of the figure of the insurance intermediary who can offer a quality supply through appropriate products—even in legal and regulatory sense—and services that don't jeopardise the customer/intermediary relationship.

The new rules of the game will bring the sectors connected to IBIPs to undergo major changes in the coming years, all aimed at a substantial alignment to the world of managed savings. If MiFID I represented a breakthrough for the financial world, tracking those changes that will be definitively completed with the entry into force of MiFID II, there is no doubt that the IDD will also represent a major breakthrough for the IBIPs market, forcing companies to revisit and heavily renovate their design and distribution strategies.

5   Conclusions

Insurance has lagged many industries in its adoption of new technology; however, this is beginning to change. An increasing number of insurers are making technology a key strategic priority.

Competition is increasing, keeping prices lower and tightening margin. To help support current income and build revenue opportunities, insurers are turning to transformative digital technologies.

Digital platforms offer the power to personalise and strengthen connections with customers with new offerings and services. Insurers also gain access to deeper insights from data analytics, and commit to new business models to better identify and reduce risk, improve segmentation, and reduce fraud. There has been a revolution in how companies can analyse data. That, combined with the new datasets from connected devices—and potentially even from social media—should enable insurers to price risk better and in different ways. We are also seeing a change in the core systems used in policy administration, claims management, and billings and payments.

Digital capabilities additionally can help meet regulatory requirements with more accurate and timely reporting, capital adequacy, and financial solvency. In addition, insurers can reduce operational costs by improving processes and increasing interactions and collaboration across the enterprise.

In-depth knowledge of current and potential markets, the development of advanced technologies, the ability to innovate in the product offering, the constant monitoring of costs and the remuneration methods of distribution channels will, in the near future, be strategic requirements for insurance industry to continue creating value in a context of continuous and rapid transformation.

References

ASIC. (2016, August). *Regulatory Guidance on Providing Digital Financial Product Advice to Retail Clients*. Regulatory Guide 255.

AXA. (2017). AXA Strategic Ventures: Placing AXA at the Heart of InsurTech.

Biener, C., Eling, M., & Schmit, J. (2013). *Regulation in Micro Insurance Markets: Principles, Practice and Directions for Future Development*. Working Papers on Risk Management and Insurance No 127.

Bughin, J., Berge, L., & Mellbye, A. (2017, February). The Case for Digital Reinvention. *McKinsey Quarterly*.

Capgemini/Efma. (2016). World Insurance Report.

Cappiello, A. (2012). *L'impresa di assicurazione. Economia, gestione, nuove regole di vigilanza*. Milano: FrancoAngeli.

Cavina, M. L., Gentile, N., & Marano, P. (2017, April). Quale futuro per la distribuzione assicurativa? *IVASS*. Quaderno n.8.

Colombini, F. (2008). *Intermediari, mercato e strumenti finanziari. Economia e integrazione.* Torino: Utet.

Craneld, A., & White, D. (2016). *The Rise of the Robo-Insurer.* Ninety Consulting Paper.

Cukier, K., & Mayer-Schenberger, V. (2013, May/June). The Rise of Big Data: How It's Changing the Way We Think About the World. *Foreign Affairs.*

Deloitte. (2015). Tech Trends 2015: An Insurance Industry Perspective.

Douady, R., Goulet, C., & Pradier, P. C. (Eds.). (2017). *Financial Regulation in the EU.* Cham: Palgrave Macmillan.

EIOPA—European Insurance and Occupational Pensions Authority. (2017, February). *Tecnical Advice on Possible Delegated Acts Concerning the Insurance Distribution Directive.* Eiopa 17/048.

Ernest Young. (2015). Global Insurance Outlook.

Guha, R., Manjunath, S., & Palepu, K. (2015). *Comparative Analysis of Machine Learning Techniques for Detecting Insurance Claims Fraud.* Wipro Research Paper.

Haddud, A., De Souza, A., Khare, A., & Lee, H. (2017). Examining Potential Benefits and Challenges Associated with the Internet of Things Integration in Supply Chains. *Journal of Manufacturing Technology Management, 28*(8), 1055–1085.

Hook, L. (2016). Generali Announces "Digital Innovation" Partnership with Microsoft.

IAIS. (2017). *FinTech Developments in the Insurance Industry.* Basel: International Association of Insurance Supervisors.

IHS Markit. (2016, May). Usage-Based Insurance Expected to Grow to 142 Million Subscribers Globally by 2023. http://press.ihs.com/press-release/automotive/usage-based-insurance-expected-grow-142-million-subscribers-globally-2023-i.

IIF—Institute of International Finance. (2016, March). Regtech in Financial Services: Technology Solutions for Compliance and Reporting.

Keller, B., & Hott, C. (2015). *Big Data, Insurance and the Expulsion from the Garden of Eden.* Geneva Association Insurance Economics, Newsletter No 72.

Lamberdon, C., Brigo, D., & Hoy, D. (2016). Impact of Robotics, RPA and Ai on the Insurance Industry: Challenges and Opportunities. *The Journal of Finacial Perspectives: Insurance, 4,* 8–20.

Lannoo, K. (2017, May). *New Market Conduct Rules for Financial Intermediaries: Will Complexity Bring Transparency?* ECMI Policy Brief No 24.

Laskowski, N. (2013, October). *Ten Big Data Case Studies in a Nutshell.* The Data Mill.

Manning, H., Bodine, K., & Bernoff, J. (2014). *Outside In: The Power of Putting Customers at the Center of Your Business.* Brilliance Corp.

McKinsey. (2010, March). The Internet of Things.

McKinsey. (2015, December). Insurance on the Threshold of Digitization: Implications for the Life and P&C Workforce.

McKinsey. (2016a). Automating the Insurance Industry.

McKinsey. (2016b). Making Digital Strategy a Reality in Insurance.

McKinsey. (2016c, March). The Economic Essentials of Digital Strategy. *McKinsey Quarterly*.

Meier, A., & Stormer, H. (2009). *eBusiness and eCommerce. Managing the Digital Value Chain*. Berlin: Springer.

Morgan Stanley. (2015). The Emerging Role of Ecosystems in Insurance.

Morgan Stanley/BCG. (2014). *Insurance and Technology: Evolution and Revolution in a Digital World*. Boston: Boston Consulting Group.

OECD—Organisation for Economic Co-operation and Development. (2017). Technology and Innovation in the Insurance Sector.

Paci, S. (2017). *Assicurazioni. Economia e gestione*. Milano: Egea.

Porter, M. E. (1985). *Competitive Advantage—Creating and Sustaining Superior Performance*. New York: The Free Press.

Porter, M. E. (1998). *Competitive Advantage—Creating and Sustaining Superior Performance*. New York: The Free Press.

Porter, M. E., & Heppelmann, J. E. (2014). How Smart, Connected Products Are Transforming Competition. *Harvard Business Review, 92*, 11–64.

Porter, M. E., & Heppelmann, J. E. (2015). How Smart, Connected Products Are Transforming Companies. *Harvard Business Review, 93*, 96–114.

Pradier, P. C., & Chneiweiss, A. (2017). The Evolution of Insurance Regulation in the EU Since 2005. In R. Raphaël Douady, C. Clément Goulet, & P. C. Pierre-Charles Pradier (Eds.), *Financial Regulation in Europe. From Resilence to Growth*. Cham: Springer.

PwC—Pricewaterhouse Cooper. (2014). Insurance 2020: The Digital Prize— Taking Customer Connection to a New Level.

Rawson, A., Duncan, E., & Jones, C. (2013, September). The Truth About Customer Experience. *Harvard Business Review*.

Rayport, J., & Sviokla, J. (1995, November). Exploiting the Virtual Value Chain. *Harvard Business Review*.

Scardovi, C. (2017). *Transformation in Insurance, in: Digital Transformation in Financial Services*. Cham: Springer.

Schmidt, R., Mohring, M., Bar, F., & Zimmermann, A. (2017). The Impact of Digitalization on Information System Design. An Explorative Case Study of Digitalization in the Insurance Business. In W. Abramowicz (Ed.), *Business Information Systems Workshop*. Springer.

Shamdami, P., Mukherjee, A., & Malhotra, N. (2008). Antecedents and Consequences of Service Quality in Consumer Evaluation of Self-Service Internet Technologies. *The Service Industries Journal, 28*, 117–138.

Swiss Re. (2014). *Digital Distribution in Insurance: A Quiet Revolution*. Sigma No 2.

Venture Scanner. (2016). Insurance Technology Market Overview–Q4.

Willis Towers Watson. (2017). How Diverse Growth Strategies Can Advance Digitisation in the Insurance Industry.

CHAPTER 3

Digital Disruption and InsurTech Start-ups: Risks and Challenges

Abstract The chapter aims to study the current expansion strategies of the new technological InsurTech start-ups, arguing that the InsurTech development should not represent a threat for traditional companies, as the digital world tools, from Big Data analysis to digital devices, from personal interactivity to home automation systems development, do not represent a direct competitive threat to the traditional activity. The chapter highlights that it is of utmost importance to remember that the digitalisation is nevertheless destined to deeply change the entire financial and insurance ecosystem. The reaction of traditional insurers will necessarily be the development of new digital competences and/or the establishment of partnerships with technology enterprises. A possible strategy is also investing in new digital start-ups.

Keywords InsurTech start-up • Insurance digitalisation • Digital disruption • Telematics in insurance • Big Data • Blockchain

1 Introduction

The insurance industry is at a critical and uncertain inflextion point. The insurance sector, considered traditional and resilient to change, is now crossed by a macro trend of digital innovation, which is bringing institutions with hundreds of years of history to rethink the insurance business

© The Author(s) 2018   29
A. Cappiello, *Technology and the Insurance Industry*,
https://doi.org/10.1007/978-3-319-74712-5_3

model, identifying which modules of their value chain to transform or reinvent through technology and data usage.

Insurers who fail to differentiate their offerings meaningfully will suffer from a lack of consumer–buyer engagement, give up business to competitors, and leave themselves vulnerable to disruptive entrants. Slow-to-adapt incumbents who insist on viewing their products as commodities, competing only on price, will not be able to succeed against those able to adopt a buyer-driven approach, learning how to attract and retain customers through brand differentiation and customer-centric capabilities.

The entry of new innovative enterprises profoundly affects the dynamics of the financial world (sphere of action of FinTech) and traditional insurance. These new competitors accelerate the transformation of the industry and drive the innovation by imposing fast reaction times. They do not merely aim to digitise the value chain, but they go further, anticipating the future needs of customers, providing intelligent services and solutions rather than pure products.

2   Digital Disruption

The insurance sector is going through a moment of significant transformation driven, among other things, by the incessant spreading of the digitisation of the industry. Insurers, traditionally rather slow innovators, began to digitise some stages of the value chain to improve the quality and speed of their offering, the transparency of operations, and the personalisation of services, and for simplifying the claims management process (Braun and Schreiber 2017).

Customers increasingly desire new ways of purchasing insurance and managing risk—including more fit-for-purpose coverage and services and more immediate delivery—and are fuelling new sales and marketing opportunities.

In the distribution, many new technologies are evolving, offering new options for consumer interaction. Distribution channels are modified by responding to changes in the preferences of consumer, who become increasingly independent in their decisions and want to interact through various channels when searching for and purchasing products insurance.

Digitisation helps insurance companies in the design of new products and the calculation of prices of new and existing ones. The growing spread of new data on insureds, collected through intelligent sensors and devices, allows for a more precise identification of the insured risks. In this regard, Internet of Things (IoT)/Big Data Analytics Technologies open the door to new ways of assessing and managing risk and claims.

The collection and analysis of Big Data will revolutionise the insurance industry. They facilitate the knowledge of potential customers and the identification of their risk profile and improve the competitiveness of products and services offered.

If, in fact, technological progress has fostered access to that precious good that is information, the Big Data management systems allow those who own them to achieve a far superior result, namely to elaborate, correlate, and analyse the data available and to draw new information and forecasts from this process, in real time and with a high level of probability. In fact, the collection and analysis of these data offer considerable opportunities regarding knowledge of potential customers, the primary objective of insurance companies as instrumental to a more efficient identification of the corresponding risk profile and an improvement in the competitiveness of the products and services offered.

In particular, the use of Big Data can guarantee an exact personalisation of the insurance offer by adapting the services, in terms of quality and price, the profile of potential customers and their specific needs.

Of course, the vast amount of data collected and the heterogeneity of the same and its sources make traditional information management tools entirely inadequate for the administration and analysis of Big Data and impose, therefore, a new challenge to operators who wish to exploit their potential. Requirements for new digital solutions are challenging legacy systems, and new technologies are offering new options for system infrastructure. The investment of considerable resources in new technologies is the means to maintain or increase the competitive advantage within their related markets.

The current context, characterised by a heightened dynamism, both about the needs expressed by consumers and about the possibilities of fulfilling their expectations, thanks to the use of advanced technologies, has seen a type of technological start-up, generically defined "InsurTech" establish itself in the insurance sector (Box 3.1).

Box 3.1   Technology Relevant to InsurTech

A number of wider technological developments and innovations underpin many of InsurTech developments. Some of the technologies are inter-related and a brief review of them is useful in establishing a common understanding of their nature.

(continued)

Box 3.1   (continued)

Mobile Technology and Applications (Apps)
The network effect of mobile phones and development of applications for these devices ("apps") have allowed many companies to reach a bigger audience than was previously possible. Mobile technology may be working in different ways for InsurTech, depending on the generation of mobile networks available and the types of handsets that are most widely used.

Smartphones and Internet access enable innovations which are based on the use of apps. For this, mobile networks that allow short messages and prepaid mobile phones, as well as large data transfers, would be necessary. This is particularly relevant to emerging markets which have low insurance penetration and do not have a well-established distribution network.

Artificial Intelligence (AI), Algorithms, and Robo-Advice
AI is intelligence exhibited by machines. A machine would be considered "intelligent" when it takes into consideration its environment and takes action to maximise the possibility of achieving its given goal. It is widely used when computer programmes are developed to have cognitive functions such as learning and problem solving. AI research is taking place in fields including reasoning, knowledge, planning, learning, natural language processing, perception, and moving/manipulating objects.

Algorithms are part of AI, where there is a set of steps for a computer programme to achieve a task under certain conditions. Well-known algorithms include route navigation systems or computer chess games. In the financial sector, algorithmic trading, such as high-frequency trading, is wide spread, with pre-programmed trading instructions to execute large trading orders. The algorithm would follow a set of conditional instructions for placing a trade order at a speed and frequency that is not possible for a human trader.

Robo-advice, or automated advice, is becoming prominent particularly for online investment and savings platforms. It can cover a broad spectrum of services, but is essentially an "on-line automated advice model that has the ability to deliver advice in a more cost-efficient way". For the insurance sector, robo-advice is being developed for investment management and is now being increasingly used for quotes with automated advice and offerings calculated through

(*continued*)

Box 3.1   (continued)

algorithms. Instead of or combined with face-to-face advice, robo-advice can provide automated guidance and execution on various financial decisions. Automated advice could assist pockets of population that do not have access to financial advice to gain input in a more cost-efficient way than a human advisor. However, depending on how the algorithm to provide advice is structured, it could also lead to inappropriate advice being made inadvertently.

Smart Contracts
Smart contract refers to any contract which is capable of executing or enforcing itself. They are written as programming code which can be run on a computer or a network of computers rather than in legal language on a printed document. This code can define strict rules and consequences that emulate a traditional legal document, stating the obligations, benefits, and penalties due to either party being in various circumstances. Smart contracts enable people to trade and do business with strangers, usually using the Internet, without the need for a large centralised authority site to act as an intermediary. The limitation of a smart contract is that a programme may not know what is happening in the physical world or react to unforeseen events, thus being unable to execute an action that was the basis of the contract.

Smart contracts often run on blockchain or distributed ledger technology (DLT). Example of a smart contract using DLT is a crypto currency, such as Bitcoin.

Blockchain/DLT
Blockchain or DLT is a protocol for the exchange of values or data over the Internet which does not require an intermediary. The protocol of blockchain technology is to create a shared, encrypted database of transactions and other information. Examples of ants and flocks of geese have been given to demonstrate what a perfect blockchain society would be like, decentralised yet coordinated. The technology is to establish an ever-lengthening chain of blocks of data. Each block has compact record of validated transaction by participants in the blockchain, and the premise of blockchain is that the information in the blocks is true. Once the transaction is validated and recorded, the stored record is irreversible. Blockchain originally referred to the database where all Bitcoin transactions are recorded and stored.

(*continued*)

Box 3.1 (continued)

Wearables
These connected devices provide valuable, near-real-time health data to insurers, helping them to better model health-related risks to create personalised pricing, speed claims processing, react quickly to health events, and foster healthy behaviour among customers.

Drones—digital imagines captured by aerial drones can speed property assessment during underwriting and loss assessment while processing claims, especially in cases such as fire or natural disaster, where the properties are difficult to access (OECD 2017).

The neologism identifies everything that is technology-driven innovation in the insurance industry: software, applications, start-ups, products, and services (OECD 2017; Drucker 2002). Borrowed from the term FinTech that refers to the more specifically financial world, the InsurTech sector is demonstrating, in the insurance sector, the same dynamic that has affected the entire industry of the financial service, with the birth and spread of start-ups that, through technology, innovate one or more steps of the value chain of traditional financial institutions (Swiss Re Institute 2017a; Carney 2017).

The entry of new innovative enterprises profoundly affects the dynamics of the financial world and traditional insurance. It is a universe punctuated by different elements: the sector includes small societies and large companies, emerging start-ups and structured businesses, in which the differences seem to exceed the contact points. Even talking about an actual market, with its borders and peculiarities, may sometimes appear improper.

All the stages of the customer journey and the value chain are affected by the InsurTech phenomenon. The macro trend of digital innovation is leading, in fact, to a much more fluid state of the sector, where each value proposition can become the integration of a set of multiple modules belonging to different players; at the same time, the boundaries become increasingly blurred between the classic roles of distributor, supplier—sometimes coming from another sector-, insurer, and reinsurer. In this scenario, the relationships of strength are questioned, and, consequently, the share of the profit pool of competence of the different participants and each one can cooperate or compete according to the context and the moment (AXA 2016).

Although the first investments in the InsurTech sector date back to 2011, there has been a significant expansion of the phenomenon since 2014 (CB Insights 2017b, c).

The number of start-ups continues to go up and, according to some estimates, will soon exceed the figure of 1500 companies, at a global level, with the USA occupying the most significant share, followed by the UK, India, and Germany (Venture Scanner 2017a, b).

In 2011, the total fund raising of the InsurTech sector was 130 million dollars, peaking at 2.7 billion in 2015, compared with about 750 million in 2014. For the future, a growth of the InsurTech platforms has been envisaged, which would bring the fund raising from 175 billion dollars in 2016 to 235 billion dollars in 2021 (CB Insights 2017d; Maynard 2017).

Most InsurTech investments are concentrated on innovations in the non-life compartment, and particularly on the health and car segments. However, the diversification by thematic is growing and if investments were primarily concentrated on marketing and distribution, which are still preponderant, they are now extended to solutions in the field of analytics and underwriting on demand, namely the specific activity of selection and risk assessment of policies sold to policyholders (Cambosu 2016; Capgemini Consulting 2015).

Many InsurTech start-ups are focusing on the role of "enablers" of internal process efficiency of insurance companies for data analysis platforms generated by the various devices. Other InsurTech start-ups, instead, are aimed at the final consumer and have wagered on brokerage services, thanks to robo-advisor, an advanced technology which mixes artificial intelligence (AI) and analytics tools.

The value chain in the insurance market is rapidly transforming, and this is forcing traditional insurers to improve their offerings and customer services in order to limit the damages deriving from the arrival of new entrants. In this regard, it is to be noted that in addition to generalist private equity companies, an increasing number of traditional insurance and reinsurance companies have begun to invest strategically in the InsurTech sector in the last two years. In addition to the investments in start-ups, some companies have attempted to keep up, configuring accelerators and business incubators, or entering into a partnership with InsurTech start-ups. The goal is to build profitable partnerships with new operators and, on the other hand, strengthen the relationship with the market. These initiatives bear witness to the fact that incumbent operators begin to understand the potential of the InsurTech sector and to consider positively the digitisation of their business model (Munich Re 2016).

3   INSURTECH START-UPS ACTIVITY

Following a rapid development of global scope, the current InsurTech panorama has become very vast and heterogeneous. The business of InsurTech start-ups does not focus on an isolated phase of the value chain, since all stages of the insurance ecosystem may be affected. For these reasons, the attempts—however small—proposed by the taxonomic systematics of the phenomenon of the literature or operators cannot represent in a unique and exhaustive way the InsurTech panorama, so extensive and liquid, where specific areas of activity assume boundaries that are not well defined and cross-border to the different categories identified (Braun and Schreiber 2017; OECD 2017; Startupbootcamp InsurTech 2015; Venture Scanner 2016; CB Insight 2016).

Below we propose a review of the main types of InsurTech start-ups present on the global market, tracing them back to three macro groups related to (1) digital customer management, (2) innovative services development, and (3) process optimisation and customer selection (Roland Berger 2017).

Digital brokers and the comparison and management platform refer to the first macro group.

Digital brokers provide insurance brokerage services through technological solutions such as online portals or mobile applications.

For a long time, traditional brokers have played an essential role in the insurance industry, supporting customers in making complex decisions. However, digital disruption is also subverting the insurance brokerage sector, as the use of technological tools allows, in a faster and more efficient way than traditional brokers, to determine the risk profile of the customer and to direct it to the most suitable insurance cover.

More and more brokerage companies are embracing digital technologies to automate their activities and improve the services offered (Mulhall et al. 2016). Digital brokerage automates the connection between brokers and insurers, insureds and potential customers, saving time and economic resources. It is able, moreover, to carry out a continuous exchange of information, to obtain informative insight, thanks to the study of data and to interact with customers through different channels.

The digital brokerage is based on four pillars that work synergistically to support the implementation of the digital strategy, providing key features to operate more efficiently, make business decisions more consciously, and develop closer customer relationships. The first pillar consists of the brokerage management system, conceived as an application software that automates the management of all business processes, from the subscription of policies to their payment. The second pillar concerns connectivity with insurers, which is fundamental to continuously and securely improve communication and data exchange, as well as to offer products that best meet the needs of customers. Mobility is the third pillar, which allows digital brokers to provide the customer with real-time information about policies through the use of smartphones. The last pillar is the use of the cloud that permits to move the primary application software to an external data centre in order to improve performance, flexibility, and security; safeguard data; and improve service to customers.

One of the first digital brokerage start-ups was "Knip," launched in Switzerland in 2013. This start-up is aimed at insureds via a mobile app. It is, therefore, a predominantly mobile service that aims to support the insured offering, in addition to the standard policies management services, also other automated consultancy services on insurance coverage.

Among the start-ups whose activity is aimed at addressing the insurance choices of the consumer, the most widespread are the aggregators, whose peculiar characteristic consists of carrying out comparison operations of a large number of insurance solutions.

Some providers have been on the market for around 20 years, so these platforms are comparatively well established. Key success factors are intuitive ease of use and the integration of relevant information relating to insurance. Although car policies are considered the product on which the price comparator sites are most concentrated (since it is a rather simple and standardised product), there are also aggregators dealing with other types of products such as health insurance or travel insurance. There are also multiproduct platforms.

Comparison platforms can begin to move forward by growing their services in the direction of an integrated comparison and management platform. Many providers combine online policy management with the services of a broker.

The aggregators are disruptive by nature. The spread of price aggregators limits, for traditional companies, direct contact with the customer, with the real danger of drastically reducing the degree of loyalty of those customers who make their choices exclusively on the basis of the economic convenience of the offer (Jubraj et al. 2016). It follows that the insurance companies that deal with or compete with the comparison portals tend to offer low-priced policies, limiting the characteristics of the products offered or reducing their quality (see Chap. 4).

The participation of the traditional insurance companies to quotation and selection services offered by the price aggregators varies from country to country. For example, in Germany, three of the top ten insurance companies, including the main one, do not cooperate with these portals, while in the UK the most important companies are directly involved with the aggregators to meet the demands and improve the interaction with online customers and reduce the cost of acquiring new customers.

The second macro group of activities InsurTech comprises different types of start-up that have developed a specialisation in convenience and customer-oriented products and services, focusing on the specific needs of specific customer segments ranging from the mobility and security segment to health insurance and asset protection. Peer-to-peer (P2P) insurance solutions can also be included in this area.

The digitisation allows for profound changes as compared with the past in the mobility and security sector. This can be referred to start-ups that have innovations focused on how to improve the sales act, such as on-demand insurance. These allow insureds to cover specific risks for a specified period. For example, travel insurance can only be activated when tickets are purchased; car insurance can just enter into force at certain times of the year in which you are using the car.

Compared to traditional coverage, on-demand insurance is more flexible, transparent, and cheaper for the costumer. Today, many companies operate successfully in this segment, providing their services also with the help of mobile apps that allow to set the insurance coverage on "on" and "off" (Braun and Schreiber 2017).

One of the first start-ups in this sector was Trōv, a Californian company founded in 2012, with the aim of reinventing the system of policies by digitising the process of purchase and customer service, making it much easier, more flexible, and transparent. Thanks to the use of high technology, this start-up proposes on-demand policies on a wide range of goods such as PCs, smartphones, TVs, bicycles, and so on. Both the activation/deactivation of the policy and the management of claims can be managed via smartphone and a live chat (Roland Berger 2017).

In October 2017, on-demand insurance has also landed in Italy with "Yolo," a start-up that proposes micro-insurance policies, also lasting for one single day, covering four sectors: travel, goods, people, and health. The model is structured on a direct channel, the platform "Yolo," and on an indirect distribution through partners. "Yolo" does not use physical channels; it is necessary to access the online platform (already active) or through the mobile app (available from 2018).

There are also many examples of InsurTech developed in the health segment. In this context, the offer is enriched with ancillary services such as health programmes, video consultations, and so on. An example is represented by Oscar Health, New York start-up, that leveraging its technology simplifies the approach to the customer through the offer of simple insurance plans. Its communication focuses on the people and their health as a good family doctor, proposing a clear offer: free check-ups, televisits, specialist advice, and a very friendly mobile application for quick information and drug research. Oscar even gives a one dollar discount per day if the client takes a certain number of steps, measured with tools such as smartwatches and fitness monitors (Agarwal et al. 2010).

A fundamental aid to the development of these insurance formulas comes from the hardware and software solutions of the IoT that represent the enabler of Connected Insurance (Gartner 2016; Jones 2014).

IoT consists of a set of devices, sensors, and other objects of everyday life connected and able to communicate with each other through the Internet. Although companies operating in this field are not InsurTech in the strict sense, their products are often employed in the insurance context.

In particular, there are four areas with high potential for insurance companies: wearables, telematics, smart home devices, and drone technology (Morgan Stanley 2014).

The examples of start-ups in this area can be numerous, which relate to various market segments. We can mention Octo Telematics, operating in the field of car insurance, which based on the latest generation telematic software can acquire data from any sensor, examine them, and provide analysis and information in real time, offering its users a fully digitised service. Another interesting example is Aerobotics, whose goal is to generate data for the agricultural, logistics, and mining industries, thanks to the exploitation of drone technology. The images taken by the drones, in fact, help the insureds in carrying out their activity and, on the other hand, allow the insurers to assess the costs of the potential claims more efficiently.

P2P insurance represents an innovative insurance formula. The P2P insurance system is very close, conceptually, to the mutualistic spirit for which all insurances were born. In fact, it brings together private parties for mutual insurance coverage. This insurance model organises insureds in purchasing groups, with the aim of getting them a saving on rates. The premium paid by the customers is divided into two portions: one part is destined for a standard insurance company, with which the company InsurTech cooperates; the other part is paid to the single fund of the group (cash back pool). First, the group undertakes to compensate any claim caused by one of the components, up to a maximum amount agreed (deductible).

For minor damages, the traditional insurance cover is not activated and the insured policyholder are firstly paid out of this common fund, while for claims above the deductible, the ordinary procedure with the insurance company is initiated. Thanks to the inclusion of the deductible, the insurance company is willing to grant a discount that can reach up to 50% of the price of the policy. The number of members allows to share and bear the cost of the possible deductible and, at the same time, to enjoy the reduction of the premium every year (IVASS 2017).

At the end of each year, the sums of the common account remaining unused are redistributed to the group's users (claims-free bonus) or reinvested into the renewal of the coverage. If the common account is zeroed or insufficient, another insurance will cover the loss, with a stop-loss mechanism. In any case, customers are never exposed beyond the initial premium.

This approach not only increases customers' satisfaction and their loyalty, but also significantly lowers the risk of moral hazard and fraudulent behaviour. In fact, the knowledge and mutual trust of the members of the group means that there is a natural disincentive to fraud.

The P2P insurance was launched in 2013 by the German company Friendsurance which introduces an innovative model of car insurance inspired by the sharing economy, developing policies aimed at small groups of friends and acquaintances (up to a maximum of 15 people) connected through Facebook.

In recent years, the P2P insurance formula has undergone an evolutionary process of its business model, which goes beyond the pure distribution model and sees its leading exponent in the US company Lemonade. This does not only work as a distributor, but it takes the risk, thus configuring itself as a real digital insurer.

The third macro sector of InsurTech includes start-ups whose activity is aimed at the process optimisation and customer selection. These start-ups, thanks to the use of the most advanced technologies, help to automate and innovate some stages and/or the entire value chain of insurance.

In this context, blockchain technology will increase the level of automation and thereby further improve process efficiency. However, this technology may not only limit itself to the process level, but may also be able to support other management aspects.

In a recent study by Boston Consulting Group (2016), it is emphasised that blockchain is destined to represent a disruptive trend for the insurance industry as it can revolutionise the way transactions are managed, linking the different parties involved in a safe and efficient way

(Evans et al. 2016). The access to secure and decentralised transactions, guaranteed by blockchain, provides an improved basis for non-repudiation, governance, fraud prevention, financial data, and reporting. Accurate and timely notification of changes drives improvements in aggregated risk and capital opportunities, as well as significant data strategies, which are founded on more available and secure information about customer assets, priorities, preferences, and third-party information services.

On the one hand, blockchain technology offers the opportunity to integrate an ecosystem of trusted third parties to reduce the costs of their global platforms, improve customer and market reach, and develop new proposition. On the other hand, it permits to enhance the enterprise governance through improved data access, third-party controls, and more sophisticated management of the risks associated with products and services (Crawford and Piesse 2016).

Not only does the InsurTech sector experience such technology, but also more traditional insurance players are investigating its potential and possible scope of application for the insurance business through investments and the formation of consortia.

An example of start-up for this category is Everledger, InsurTech that started its business by offering the DLT for diamond owners. This start-up is currently part of the Hyperledger community, committed to promoting open source collaboration in the blockchain area and has succeeded in intercepting the high demand for traceability and anti-counterfeiting by the luxury market, which makes the authenticity of the products an essential added value.

Taking into account their business model, insurance companies typically manage complete databases that can be exploited to identify target customers, calculate rewards, reduce claim costs, detect fraudulent behaviours, and continuously assess the level of risk of the company. However, due to legacy systems in the insurance sector, data are often stored in a decentralised way, so it is difficult to access all the necessary information and perform significant analysis in a short time.

The InsurTech start-ups operating in the Big Data Analytics sector offer software solutions that structure and analyse large volumes of data of various kinds and from different sources, both internal and external. The Big Data, of course, constitute the access key to identify each customer, not only from the demographic point of view but also from the behavioural one. In this context, the insurance offer is transformed into a consultancy service focused on listening to and mapping the real needs of the consumer, aimed at creating highly personalised policies. This also permits to limit fraud and generate competitive advantages.

At the moment the number of start-ups operating in this sector is limited, but their potential is highly innovative. Synerscope gives an example of this category of start-up: it is a Dutch InsurTech, already working successfully on the market. Synerscope processes both structured data (database) and unstructured data (text, video, and audio files) through recognition algorithms and displays existing interrelations. This technology can be used to identify fraud, to conduct an in-depth analysis of losses in less profitable business branches, or finally to identify possible risk aggravations and the consequent adjustment of the premium.

Unlike the data analytics and various robotic process automation solutions, the use of AI is able to revolutionise the entire insurance landscape since it allows you to design and offer more and more services tailored to the needs and expectations of the consumer, digitising the whole chain of the value of the insurance process: from product design to sale, from the subscription of the policy to customer service to the management of claims (Acord and Surely 2016; Stuart 2013).

These InsurTech start-ups adopt very flexible AI solutions, without having to collide with a previous legacy of pre-existing computer systems and strict organisational structures, as is the case for the incumbent. In the light of digital experience and transformation in other sectors, it can be said that in the medium/long term the digitisation, more or less extensive, of the production process will still be a mandatory step, also for traditional insurers.

The current existence of InsurTech in this area is not yet very high, but a wide spread is to be expected shortly. A sample is given by Lemonade, available at the moment only in New York and just for household and property insurance.

4   Is the Insurance Industry Disrupted?

For many years, the insurance sector has been reluctant to change, proving to be one of the most conservative sectors. This was also due to the entry barriers represented by a strict regulation, the complexity of the products offered, and large capital requirements.

However, the innovative business model of the InsurTech start-ups is raising increasing concerns about their ability to constitute a threat for the incumbent companies, due to the process called "digital disruption" (Braun and Schreiber 2017).

The term "disruption" comes from the concept of disruptive innovation, coined in order to try and identify the effect of the introduction of a new technology into a shared business model. Bower and Christensen distinguish sustaining innovations and disruptive innovations, highlighting the different impact they have on the sector of reference (Bower and Christensen 1995).

The sustaining innovations only create an incremental advantage for the customer, as they tend to make use of competences and knowledge, which are already an integral part of the company and aim at improving the existing products performance, thanks to the introduction of additional features and functions, which often exceed the consumers' expectations (King and Baatartogtokh 2015). On the other hand, disruptive innovations have the ability to create a new market or to rewrite the features of an already existing market, presenting a new product or a new business model. In most cases, the companies implementing this type of innovation are small and have a strong risk appetite and limited resources. These companies address the lower part of an existing market or aim at a completely new market, later tending to acquire new market shares, until they have expelled the incumbents from the market (Christensen et al. 2015). The most relevant element is maybe their ability to anticipate customers' future needs, as compared to incumbents, who could be unable to innovate their business model in a timely manner, as they cannot overcome the trade-off between innovation and their current business success (King and Baatartogtokh 2015).

It is now necessary to answer the question whether and to what extent the insurance industry is vulnerable to a large-scale disruption, caused by the entrance of the new InsurTech competitors (Libert et al. 2017; Morgan Stanley 2016).

It is necessary to point out that the InsurTech sector has various degrees of maturity. Next to more structured segments, there are areas which are still undergoing a development and consolidation phase. The first generation of InsurTech has focused on the ecosystem sectors which are poorly regulated or only require a little knowledge of insurance techniques, such as the contact with the customer. However, several other scenarios can be envisaged for the near future: there are many hypotheses and some are not easily predictable. In the sole field of connected insurance, for example, projections range from home insurance to the health sector, up to the agriculture sector. Initially focused on the sole personal lines segment, InsurTechs are showing an increasing interest towards the world of small enterprises and the life sector, which had attracted less interest at the beginning. This can be seen, for example, by observing the US market, where the InsurTechs are also growing in the segment of pure life-risk policies, which presents a relevant magnitude and a high level of maturity.

As a first approximation, we can state that the expansion of investments in technology of recent years, on the one hand, and the spread of the use of technologies among the public, on the other hand, greatly increase the probability of survival and expansion of the InsurTech segment.

The context seems to favour a further growth of the InsurTech, despite the high level of regulation characterising the insurance sector. Conversely, the very presence of entry barriers hinders the access to the market of digital world giants, such as Google, Facebook, or Amazon. Without players in a strongly dominant position, the sector has the possibility to gradually and widely develop, giving the InsurTech the time and room for manoeuvre to gather funding and develop new solutions.

However, the difficulties that a start-up may encounter are manifold, so that some tech-led initiatives in insurance will inevitably fail. Factors of disadvantage are due to a poor knowledge of the market, the lack of an appropriate business model, as well as the high level of competition in the insurance sector, characterised by many complexities and a high level of technical content (Celent 2017; Fitzgerald 2017; Dietz et al. 2016; Fitzgerald and Macgregor 2016). In fact, if the new players generally have strong competences in terms of customer experience, simplification, and speed of processes, traditional companies have an important advantage, as compared to the competitors entering the sector, which is the huge pool of information about the customers, in terms of biographical data and, above all, risk profiling.

Recent surveys also report that customers do not seem ready to abandon the traditional insurance providers, as they consider them more reliable in terms of safety and protection against frauds, attaching a great value to the brand reputation and to the personal interaction (Capgemini and Efma 2017).

InsurTech and BigTech do not pose, therefore, an immediate competitive threat to established insurers. A drastic disintermediation of the insurance companies, which would also imply a deep innovation of the business models carried out by the incumbents, does not seem to lie ahead in the short–medium term.

The majority of digital competitors, at least for the moment, are not in a disruptive position, according to Christensen's theory (Christensen et al. 2015;

Braun and Schreiber 2017). They can rather be considered enablers, who are able, thanks to their technological competences, to facilitate and make the traditional insurance business more efficient.

Historically, most innovation in insurance tends to happen incrementally, influenced by gradual shifts in consumer behaviour, risk-absorbing capabilities, and the regulatory framework. Nevertheless, insurers do need to keep on top of developments because over time, future market entrants could build on the infrastructure currently being created to offer new risk-protection solutions, which can have a strong disruptive nature.

An increasing number of insurers now regard investment in digitalisation as a priority, especially considering the sector has lagged behind its financial services peers in adopting digital technologies owing to regulations, reluctance, and costs (Willis Towers Watson 2017; Naylor 2016, 2017).

Many incumbent insurers are seeking to upgrade their digital capabilities, especially in order to boost customer engagement and collect data about new risk pools. In some cases, insurers have increased spending on research and development to foster innovation in-house. Some are working with BigTech, while other insurers are investing directly in and/or partnering with start-ups. Furthermore, the majority of the entrants also seems to be willing to adopt a collaborative strategy with incumbent companies (AXA 2017; PWC 2016; Bartoletti 2014).

The development of alliances with the new competitors (such as InsurTech suppliers) allows the incumbents to take advantage of the competences, dynamics, and ways of doing business which, for its very nature, the insurance industry could not have developed. Big Data Analysis and blockchain projects are now the most interesting developing areas in the medium term for the insurance sector (Braun and Schreiber 2017).

If insurance has, as its core business, the risk selection and the price determination to take them, the current evolution towards a full usage of Big Data will radically alter the data type itself, the way they will be analysed, the claim management, and the relationship with customers.

The growing proliferation of data about insureds, be it collected via dedicated sensors, smart mobiles, or other devices, provides an opportunity for more granular underwriting of individual risks. Smart analytics, predictive modelling, and connected telematics devices assist insurers in designing products and setting premiums based on how insureds actually behave, rather than using general proxies (Schanz 2015).

Technological advances will change the degree of asymmetric information that often characterises insurance markets. Companies with innovative pricing models and information on individual risks can better identify the lowest risk clients, while self-informed, higher risk clients may seek out less sophisticated providers offering more attractive rates, based on less information. In this environment, late adopters of new technology would be more susceptible to the threat of adverse selection.

Insurers also recognise the challenge and opportunity to leverage digitalisation to create operational efficiencies throughout the business that will not only manage costs and risks, but also streamline processes to enhance customer experience.

Digital distribution offers opportunities to find new ways to market and to build closer relationships with consumers (Kotler and Miller 2014). Insurers are now called to design digital infrastructures, which are able to improve customer engagement through distributive channels that are likely to be addressed through a combination of internally driven innovation, joint ventures, and Merger & Acquisition activity (McKinsey 2016a, b; Deloitte 2015; Hirt and Willmott 2014). Insurers that hesitate could very well get left behind and fail to capture future generations of younger policyholders, who are more likely to engage via digital channels.

5   Conclusions

A series of technology changes and adoptions, many of which are interconnected, is having a significant impact on the insurance sector. They are profoundly changing the strategic context: altering the structure of competition, the conduct of business, and, ultimately, performance across industries. Digitisation often lowers entry barriers, causing long-established boundaries between sectors to tumble.

This complex scenario leads to the identification of the digital innovation as a necessity, and only when this necessity is considered an opportunity, it will be able to create value for the insurance industry as a whole.

The relatively slowness of the insurance sector in adopting these technologies is mainly due to a cultural resistance, which will have to be overcome in order for the sector to maintain its competitiveness. The organisational systems that are rigid and unable to take full advantage of the possibilities connected to the digitalisation and the Big Data are condemned to a competitive decline.

The insurers' investments in InsurTech start-ups can contribute to stimulate the innovation, identify the priorities, and integrate the digital strategies, which have already been initiated (Deloitte 2014a, b).

However, in the short term, a number of hurdles must be overcome. In particular, insurers adopting new technology often face constraints from poorly integrated IT systems and technical skills gaps. Cumbersome legacy technology limits their ability to rapidly launch new products and react to the change. The current model also typically delivers infrequent consumer interactions and often has inherent channel conflict (e.g. agency versus direct channels) (Genpact Research Institute and ACORD 2016).

Some studies suggest that, in the short term, digital technology may actually diminish revenues. Moreover, the development of multiple distribution channels likely adds to insurers' costs, at least in the short term. This reflects not only the higher operational complexity of managing many more channels and possible points of customer interaction, but also the additional upfront costs of, for example, establishing a strong brand that is often essential for direct, e-commerce sales.

This means that insurers must focus on boosting productivity across all distribution channels.

Smart analytics can help by allowing more informed monitoring of the efficiency of both traditional and new digital channels. Similarly, sophisticated AI-led automation systems (e.g. chatbots) can help remove unnecessary costs and improve the productivity of new and existing insurance intermediaries.

The transition will not necessarily be easy. However, success requires a considerable effort, which also implies a deep change able to develop comprehensive digital strategies affecting all elements in the insurance value chain to remain competitive and profitable.

REFERENCES

Acord & Surely. (2016, February). AI—The Potential for Automated Advisory in the Insurance Industry.

Agarwal, R., Guodong, G., DesRoches, C., & Jha, A. K. (2010). Research Commentary—eDigital Transformation of Healthcare: Current Status and the Road Ahead, Inform. *Systems Research, 21*(4), 796–809.

AXA. (2016, October). Verso le Assicurazioni 4.0? Il settore assicurativo e la rivoluzione dei dati, Italian Axa Paper, no.8.

AXA. (2017). AXA Strategic Ventures: Placing AXA at the Heart of InsurTech.

Bartoletti, D. (2014, March 6). *Strategic Benchmarks 2014: Server Virtualization*. Forrester Research Inc.

Boston Consulting Group. (2016). Digital Disruption in the US Small-Business Insurance Market.

Bower, J. L., & Christensen, C. M. (1995). Disruptive Technologies: Catching the Wave. *Harvard Business Review, 73*(1), 43–53.

Braun, A., & Schreiber, F. (2017). *The Current InsurTech Landscape: Business Models and Disruptive Potential*. St. Gallen: Institute of Insurance Economics I. VW-HSG, University of St. Gallen.

Cambosu, D. (2016). Insurance Tech, All Incubators and Specialized Accelerators, Insuranceup.

Capgemini and Efma. (2017). World Insurance Report 2017.

Capgemini Consulting. (2015, February 7). Strategies for the Age of Digital Disruption. *Digital Transformation Review*.

Carney, M. (2017, January). *The Promise of FinTech. Something New Under the Sun?* Speech Given to the Deutsche Bundesbank G20 Conference on Digitising Finance, Financial Inclusion and Financial Literacy.

CB Insights. (2016, July). Analyzing the Insurance Tech Investment Landscape.

CB Insights. (2017b, January). Where Insurers and Reinsurers Invested in Tech Start-ups in 2016. Retrieved from www.cbinsights.com/blog/2016-insurance-cvc-total/.

CB Insights. (2017c). The Top Financings, Partnerships & Hirings in Insurance Tech–Q1.

CB Insights. (2017d). Where Insurers and Reinsurers Invested in Tech Startups in 2016.

Celent. (2017). Success Factors for InsurTech/Incumbent Partnerships.

Christensen, C. M., Raynor, M., & McDonald, R. (2015). What Is Disruptive Innovation? *Harvard Business Review, 93*(12), 44–53.

Crawford, S., & Piesse, D. (2016). *Blockchain Technology as a Platform for Digitization. Implications for the Insurance Industry*. London: Ernst Young.

Deloitte. (2014a, January). Big Demands and High Expectations: The Deloitte Millennial Survey — Executive Summary.

Deloitte. (2014b, February). The Omnichannel Opportunity: Unlocking the Power of the Connected Consumer.

Deloitte. (2015). Tech Trends 2015: The Fusion of Business and IT. An Insurance Industry Perspective.

Dietz, M., Khanna, S., Olanrewaju, T., & Rajgopal, K. (2016). Cutting Through the Noise Around Financial Technology. *McKinsey Review*.

Drucker, P. F. (2002, August). The Discipline of Innovation. *Harvard Business Review*.

Evans, P., Aré, L., Forth, P., Harlé, N., & Portincaso, M. (2016, December). *Thinking Outside the Block. A Strategic Perspective on Blockchain and Digital Tokers*. BCG Publications.

Fitzgerald, M. (2017). *Success Factors for InsurTech/Incumbent Partnerships*. Celent, USA.

Fitzgerald, M., & Macgregor, J. (2016). *Insurer-Startup Partnerships: How to Maximize Insurtech Investments*. Boston: Celent.

Gartner. (2016, April). *Measuring the Strategic Value of the Internet of Things for Industries*.

Genpact Research Institute and ACORD. (2016, November). *Assessing Digital Impact Across Insurer and Channel Operations*.

Hirt, M., & Willmott, P. (2014, May). Strategic Principles for Competing in the Digital Age. *McKinsey Quaterrly*.

IVASS. (2017, April). La distribuzione assicurativa. Quaderno no.8.

Jones, N. (2014, April). The Internet of Things Will Demand New Application Architectures, Skills and Tools. *Gartner*.

Jubraj, R., Thomas, T., & Sandquist, E. J. (2016). Coming to Terms with Insurance Aggregators: Global Lessons for Carriers, Accenture Strategies.

King, A. A., & Baatartogtokh, B. (2015). How Useful Is the Theory of Disruptive Innovation?

Kotler, P., & Miller, K. (2014). *Marketing Management* (14th ed.). Upper Saddle River: Prentice Hall.

Libert, B., Beck, M., & Wind, J. (2017, March). How Insurers Can Protect Against Digital Disruption. *Knowledge@Wharton*.

Maynard, N. (2017, December). *Fintech Futures: Market Disruption, Leading Innovators & Emerging Opportunities 2017–2022*. Juniper Research.

McKinsey. (2016a, May). An Incumbent's Guide to Digital Disruption. *McKinsey Quarterly*.

McKinsey. (2016b). Making Digital Strategy a Reality in Insurance.

Morgan Stanley. (2016). North America Insight: Digital Disruption in Small Business Insurance.

Morgan Stanley/BCG. (2014). *Insurance and Technology: Evolution and Revolution in a Digital World*. Boston: Boston Consulting Group.

Mulhall, J., Chauhan, A., Lindsey, C., & Lyman, M. (2016). The Broker of the Future: Winning in a Disruptive Environment, Accenture Strategies.

Munich Re. (2016, January). Reinventing Insurance for the Digital Generation.

Naylor, M. (2016). A Perfect Storm in Insurance: How to Survive the Looming Waves of Disruptive Technology.

Naylor, M. (2017). The Response of Incumbents. In *Insurance Transformed*. Palgrave Studies in Financial Services Technology. Cham: Palgrave Macmillan.

OECD—Organisation for Economic Co-operation and Development. (2017). Technology and Innovation in the Insurance Sector.

PWC. (2016). How InsurTech Is Reshaping Insurance.

Roland Berger. (2017, June). Copy Them? Work with Them? or Buy Them? *InsurTechs and the Digitalization of Insurance*.

Schanz, K. U. (2015, May). The Technology and Data Revolution in Insurance: A Brave New World? *Middle East Insurance Review*.

Startupbootcamp InsurTech. (2015). So, What Is an InsurTech Startup?

Stuart, R. (2013, October). *Analytics and the Analytical Insurer*. Society of Actuaries.

Swiss Re Institute. (2017a). Technology and Insurance: Themes and Challenges, June.

Swiss Re Institute. (2017b, May 3). World Insurance in 2016, Sigma.

Venture Scanner. (2016). Insurance Technology Market Overview–Q4.

Venture Scanner. (2017a, September). Startup Market Reports and Data.

Venture Scanner. (2017b). Where in the World Are Insurance Technology Startups?

Willis Towers Watson. (2017, January). How Diverse Growth Strategies Can Advance Digitisation in the Insurance Industry.

CHAPTER 4

The New Frontiers of Insurance Distribution

Abstract The chapter focuses on the role of the distributive variable in the insurance industry, and the delivery process, composed of the sales staff, clients, and physical support. According to the higher or lower incidence of the three mentioned variables, as well as the specific expectations and needs of the various market segments, different delivery systems are envisaged. The choices regarding the planning of differentiated delivery systems are therefore strictly linked to the analysis of the peculiarities and the competitive dynamics of the strategic business areas where the company operates. These business areas are defined on the basis of three key elements, that is customer segments, the needs they express (service types), and the ways these needs are met (productive-distributive technology). In this regard the most innovative technological distributive channels are analysed, by examining in-depth risks and opportunities.

Keywords Insurance delivery system • Servuction • Service management system • Innovative insurance distribution

1 INTRODUCTION

The distribution variable assumes a fundamental role in the insurance industry field, where the intangibility of the supply ensures that there is a close link between the insurance company on the one side, and the customer on the other, within the delivery process of the service, consisting

© The Author(s) 2018
A. Cappiello, *Technology and the Insurance Industry*,
https://doi.org/10.1007/978-3-319-74712-5_4

of human factor, customers, and physical support. In this way, different delivery systems are configured according to the greater or lesser incidence of the three variables mentioned and the specific expectations and requirements of the various market segments.

It is, therefore, necessary to plan an offer system that involves the diversification of distribution channels and delivery systems according to the different services offered and to the different segments of customers to be reached.

The choices about the design of differentiated delivery systems are therefore closely linked to the analysis of the peculiarities and the competitive dynamics related to the strategic business areas of where the company is to operate. These business areas are defined based on three key elements such as customer segments, the needs these expressed (types of services), and the ways for satisfying these needs (production-distributive technology).

2   THE INSURANCE SERVICE AND ITS DISTRIBUTION: MAIN FEATURES

The insurance product and its distribution assume specific connotations for a twofold set of reasons, the one due to the peculiarity of the insurance activity and a relationship of trust that is established between the customer and the insurance company and the other relating to the service nature of the insurance output (Normann 2001).

With regards to the first aspect, it suffices to point out that the object of the insurance activity assumes that the external environment projects expectations related to the achievement of security and allocation efficiency objectives onto the company and the insurance system overall.

The insurance relationship also implies a reiteration of the contact between the two contracting parties, which is founded solely on mutual reliability judgements. The customer relationship, therefore, assumes, in this regard, connotations of greater complexity than any other kind of company.

Furthermore, the intrinsic characteristics of the insurance product must not be neglected, firstly pertaining its immateriality, from which the features of non-storaged, non-transportability, and non-verifiability also derive, which directly influence the way the product is delivered (Normann 2001; Eiglier and Langeard 2003).

The lack of material requirements constitutes the first distinctive element of the insurance service. The user's assessment criteria cannot, therefore, relate to the service as such but must, rather, be based on the objective elements which represent the technical support for its use, such as the human factor (Kotler and Pfoertsch 2011).

The intangibility of the service also affects the production and distribution process. In the field of insurance management, unlike in industrial companies, the production stage cannot take place before that of the sale; the production process is, in fact, activated by the specific request of the customer. It is very clear, in this regard, how a strong interaction between the delivery unit and the demand unit is created (Black et al. 2002).

This event then assumes greater significance for the fact that the insurance service, being intangible, has no autonomous identity from the manufacturer, and therefore must be provided directly to the customer who requests it. It follows that, in most cases, the stages of production, sale, distribution, and consumption have a high degree of interconnection and are not easily identifiable in an autonomous manner. This means a very close relationship between the insurer and the customer that underlies a direct and repeated contact between the two parties.

Moreover, the intangibility of the service means that this, in general, cannot be stored, or conserved in any way, by the company that produces it, nor by the customer. Regarding the latter, it seems obvious that this circumstance requires a more frequent relationship with the dealer, whether it is traditional type (agency) or innovative type (technological applications of various kinds).

On the other hand, the company cannot store the "finished product" and must necessarily provide for the design of an offer system capable of dealing with turnout peaks, keeping waiting times within acceptable limits; the speed of operations constitutes, in fact, a factor that is not negligible in the choices of the public (Kotler 2001).

In order to contain the size of the delivery structure, the insurance company can obtain a lowering of the demand peaks by delegating part of the delivery process to the customer through the dissemination of digitised techniques. In this way, given the possibility of accessing the service at any time during the day, and in places other than the physical structure, a part of the demand for services of simpler enjoyment decreases at the traditional point of sale.

It is also to note that the intangibility of the service determines, in general, the "non-transportability" of the same. This means that the service must be produced, delivered, and consumed at the offer unit, since it cannot assume autonomous identity from the company producing it. As the mobility of users cannot be regarded as unlimited, the provision of an adequate distribution network is essential to allow and facilitate the meeting between the company and its market.

If the above can be agreed with in its general lines, it is necessary to point out, more specifically, that the insurance product can be considered to be transferable in space to the extent that it is possible to separate the delivery process into a stage of contact and a stage which is to be carried out by the customer at a distance. In fact, the presence of the customer becomes necessary at the time of sale and often in the distribution, whereas the stages of production do not necessarily require the intervention of the customer itself.

This means that the different "processing stages" prior to obtaining the "finished product," and which do not involve direct contact with customers, can also be managed at considerable distance from the latter, according to operational efficiency objectives.

It is possible to hypothesise organisational solutions envisaging the production of the service in legal entities other than those involved in the distribution of the same. This circumstance may seem to contradict the theoretical need to configure direct channels for the distribution of the insurance service, based on the intrinsic peculiarities of the same. However, it is not to exclude the possibility of configuring indirect solutions, in case it is necessary to delegate the function of service selling to third parties, which does not have any dissimilar characteristics from the marketing of any other good. It seems plausible, moreover, to refer to indirect channels also for the physical distribution of the service when the producer delegates some stages of the production process, and, in particular, the phase of creation of the finished product.

The non-demonstrability of the service in tangible terms, finally, means that the benefits related to it may seem difficult to understand by the customer. The service, in fact, assumes the character of the concreteness only at the time of its use; in that moment only, the customer can evaluate its usefulness and quality (Normann 2001). If we take account of the fact that, once it has been delivered, and therefore consumed by the customer, the service cannot be returned, it is understandable that the demand unit can assume attitudes of distrust and mental reserve towards the providing company.

In this regard, the importance of the human factor in the production/ distribution process of the insurance service is clear. Given, in fact, the peculiarities of the production process of the latter, which underlies a close link between the insurer and the client, the intermediary in contact with the customer must assume, at the same time, technical and commercial functions. In many cases, the intervention of specialised human tellers able to provide, through a comprehensive range of information, a more "concrete" connotation to the service offered will be crucial.

3   THE INSURANCE DELIVERY SYSTEM

From the foregoing, it is apparent that the process of design and distributing the insurance service presents completely peculiar connotations that deeply affect the market strategies of the insurance companies.

Of this type of company, in fact, unlike what happens in industrial companies, the delivery of the service in the broad sense, defined *servuction* (Eiglier and Langeard 2003), comes to be configured in a process that sees the phases of production, sale, distribution, and consumption as closely connected, whose independence and overlapping depend essentially on the type of service delivered.

Hence, the distribution of the service implies a close interaction between producer (insurer) and consumer, which becomes an integral part of the process of delivery itself. It is true that during the stages of acquisition and use of the insurance service, more or less frequent contacts are needed between the customer and the insurer (Dumm and Hoyt 2003).

The strong interaction between the insurance company and the client is also underlined by the fact that the service, even when it enters the sphere of availability of the user in the initial stage of delivery, does not come out of the sphere of availability of the manufacturer and turns out to be continually influenced and activated by the latter (Normann 2001). In fact, the company can modify the enjoyment manner of the service in any time with its action (e.g. with the variation of the premium or certain conditions). On the other hand, the user with his conduct can influence the delivery and therefore the use of the service (e.g. in the case of increase in risk or moral hazard behaviour).

Therefore, an interactive process is created between the two counterparties that together contribute to the creation and maintenance of the insurance relationship.

As the user is called to participate actively in some phases of the delivery process, the time of contact between the customer and the company (the "Moment of Truth") assumes, for the company as for any other service company, a very special meaning and a strategic value (Eiglier and Langeard 2003). Only in the fruition phase the customer can evaluate the service itself and judge its quality.

The very nature of the insurance product, therefore, and the consequent interrelation between demand and supply ensure that the service delivery system represents one of the critical factors of success of company strategies as a key element of the more general service management system.

The service delivery system is therefore composed of the following elements: (1) agents and brokers, (2) customers, and (3) physical support.

Of the identified elements, agents and brokers can be considered the most important. Together with the physical support, it is part of those tangible elements that can determine the image of the company vis-à-vis the customer. The latter, in fact, from the observation of the behaviour of the human factor draws elements of judgement on the efficiency, the speed, the courtesy, and therefore the quality level that distinguishes the relationship with the manufacturer (Su Chen and Lai 2010).

The customer, on the other hand, represents the second fundamental element of the delivery system; the direct involvement of the consumer is in fact crucial for the completion of the *servuction*. This highlights the importance of this factor, precisely because of the close, repeatedly recalled, correlation between "producer" and "consumer."

It is obvious that even marketing decisions cannot overlook this circumstance, by placing the human element (both the sales factor and the customer) at the heart of the different company policies. To increase the degree of customer satisfaction, it is advisable to promote a greater and more active participation of these at the various stages of the production process.

Of course, even the other customers who are at the place and at the time of use of the service help to connote the delivery system of the same. They may represent a reason for satisfaction or annoyance for the individual customer; this may depend on the type of interrelations created in the waiting areas, from the presence of people belonging to more or less homogeneous market segments, and so on (Thornton and White 2001; Su Chen and Chang 2010).

Finally, the physical support includes both the environment where the service is delivered, and the technological support needed for its production and distribution.

As we can see, physical support considerably contributes to activating and improving the interaction between the company and the customer. The place of use of the service, for example, as a meeting point between the unit of supply and the units of demand for an intangible product, becomes the primary source of contact and communication of the company with the market.

The convenience of access to the "distribution point," given the impossibility to transport or storage the service, constitutes an element of great significance within the customers' evaluation system, since it allows them to reduce costs, both material and psychological, linked to the enjoyment of the service itself.

The localisation of the delivery units also acquires relevance; in fact, the more the service is "near," both in a spatial and temporal sense, to the customer, the more these will be inclined to its use.

In this regard, Information Technologies (ITs) play a leading role in the "production" of services, allowing a smoother standardisation of their quality level and a better interaction with the customer even in places other than the traditional point of sale. Since technological innovation can clearly contribute to the differentiation and innovation of the production process, and therefore of the insurance service, it is clear that it should be considered a strategic variable of undoubted efficacy (Allmendinger and Lombreglia 2005; Brady et al. 2005).

The three elements we have briefly examined, namely the agents/brokers, the customers, and the physical support, combined, concur to characterise the service and to determine its qualitative level; they also constitute a differentiation factor of the delivery system, and therefore must be managed according to the expectations and needs of the various market segments.

In this way, different delivery systems are configured according to the greater or lesser incidents of the three variables mentioned and the interrelationships that come to be established. It is clear that different distribution channels also imply different delivery systems (think of the traditional agency opposed to the digitised distribution solutions). On the other hand, it is also clear that a single distribution channel can relate to multiple delivery systems activated simultaneously; at the agency, for example, various delivery systems can coexist to offer products of different nature and to serve non-homogeneous demand segments (Easingwood and Storey 2006; Coelho and Easingwood 2005; Trigo-gamarra and Growitsch 2010).

It is, therefore, necessary to plan a delivery system that involves the diversification of distribution channels and delivery systems according to the different services offered and to the different customer segments to be reached.

4   TECHNOLOGICAL INNOVATION AND EVOLUTION OF DISTRIBUTION CHANNELS

The dynamism of the competitive scenarios leads to the adoption of policies of differentiation and innovation of the services offered, in which the realisation of appropriate distribution solutions assumes a critical role.

The intangibility of the insurance service makes it complex to use the product lever to acquire stable competitive advantages. Suffice it to remember that insurance product, as it is intangible, cannot be the subject of differentiation policies that affect the extrinsic character of the same; furthermore due to the lack of physical requirements, this product can be imitated without the possibility of patenting it.

The peculiarity of the insurance business requires, therefore, the use of appropriate methods of differentiation of the products/services, among which the distribution variable constitutes, without doubt, a valid instrument available to the companies (Fürst et al. 2017; Klotzki et al. 2017).

In fact, the interdependence between the production, distribution, and consumption phases makes the delivery system decisive in the connotation of the service, as well as in its qualification with the customer.

The combination of the components of the delivery system, that is, the physical support, the agents/brokers, and the customer, and the relationships that come to be established between these elements are therefore strategically important.

The differentiation of the product can be implemented through a change of technological, relational, and organisational kind, which interests the constituent elements of the delivery system, and consequently modifies the use of the product by the client (Nightingale 2003; Pires et al. 2008).

In this sense, new opportunities for innovation in the insurance business are also offered. It is evident that technological evolution acquires a critical character in the renewal of the production and distribution processes of insurance services, as well as of their innovation (Coelho et al. 2003).

The adoption of advanced technologies, in fact, in addition to contributing in an incisive way to the rationalisation of the systems of delivery of insurance services, also promotes the possibilities of differentiation and innovation of the same (Brown and Goolsbee 2012). More specifically, the change introduced by technology may concern the product/service in the strict sense—reference will be made, then, to the introduction of new insurance services—or the production and distribution process of the product/service.

In this case, thanks to technological progress, the relationships between the different elements of the traditional delivery system are sometimes also considerably modified. For example, the subscription and use of an insurance policy which can be carried out alternately with the agency, or through digitalised solutions, is thought to be the case. It is apparent that the purchase place of the service, the distribution procedures, and the insurer–client interaction change considerably from one hypothesis to another, modifying the qualitative and functional characteristics of the service.

It is useful to underline, in this respect, that in the field of insurance activity, there is not always a clear distinction between product innovation and process innovation. Since the "finished product" and the different phases of its delivery system are often strictly combined, it is not difficult to detect how the differentiation and innovation of the production and distribution process also profoundly affect the qualitative and functional characteristics of the service becoming a peculiar element. The different technological solutions, therefore, sensibly innovating the delivery process, also affect the degree of novelty of the service itself.

In the light of the considerations on the subject, we want to emphasise the importance of IT not only as a tool for streamlining procedures and saving costs to these related, but also, and above all, as a factor of differentiation and innovation of the product/service and its delivery process, for the achievement of stable competitive advantages (Kabadayi et al. 2017).

In fact, the growing dissemination of IT allows for the expansion of distribution channels, among which are the ones of technological nature, while, on the other hand, promotes the transformation of the agency both regarding its physical and functional configuration, and in relation to its role within the relationship with the market.

It should not be neglected, in this respect, that the expansion of operational boundaries, financial innovation, and the changing needs of the market have led to the emergence of the favourable conditions for diversification, as well as for a certain specialisation, of the delivery systems (Heinhuis and de Vries 2009).

As is obvious, this tendency leads to a careful assessment of the problems inherent in the composition and coordination of the entire distribution system related to the products offered and to the market segments served.

However, it is useful to underline that these distribution channels have different functional characteristics that depend directly on the type of service offered and the peculiarities of the market segments served. The different nature of the needs expressed in the various market segments, and the consequent specificity of the services offered, therefore, require special methods of delivery, which relate to certain distribution channels (Coelho and Easingwood 2008).

The choices for the design of differentiated delivery systems are therefore closely linked to the analysis of the peculiarities and the competitive dynamics related to the strategic business areas of where the company is to operate. These areas of business are defined by three key elements such as (1) customer segments; (2) the needs they expressed (types of services); and (3) the ways for satisfying these needs (production-distributive technology).

In this regard, it is worth mentioning that insurance services are placed between an extreme of simplicity and low unit added value, and an extreme of high complexity and significant added value; in this regard, elementary services and complex services are identified.

To the first category belong all those services which, characterised by the execution of simple operations, have a minimum content of personalisation, and are easily standardised (e.g. motor insurance, home insurance, etc.).

For the distribution of such services, whose demand is determined, in particular, by the price and comfort, understood by the latter as proximity of the point of delivery and speed of execution of transactions, the insurance company may have systems of relatively inexpensive delivery, and based on the use of advanced technologies.

The use of technology, allowing the standardisation and capillary distribution systems, permits to achieve advantages of cost structure (and to develop, consequently, competitive strategies based on price) and to respond effectively to customer expectations to increase the distribution of sales points, to reduce waiting times, and to expand service hours (Black et al. 2002; Thornton and White 2009).

The complex, or specialist, services, on the other hand, have opposite characteristics: in summary, they require a high degree of personalisation and are of high added value (e.g. life insurance and pension products). The customers, in this case, are sensitive to the quality of the service offered and to the personal relationship since these services require some assistance and consultancy services before and after the sale in the strict sense.

It is not possible, in this specific case, to transfer a large part of the executive operations to automatic devices, as it detects the degree of "personalisation" of the relationship; the company must have delivery systems that allow to develop an unstructured, high-content, "service" relationship. For the distribution of services of complex type, for which a high level of interaction with the customer is required, it is, therefore, necessary to provide channels with high-quality characteristics, strong personalisation, and operational flexibility. It is conceivable in this regard a delivery system that implies a good level of contact with the customer; this can be done at the agency, or through dedicated delivery networks.

It follows that the complexity of the delivery system tends to increase whenever the complexity of the service provided and of the relationship with the customers increases. In general, it can therefore be said that the most complex delivery systems, and with higher added value, are dedicated to the distribution of product services that have a higher specialist content, such as life policies with connotations of more or less high investment, while the delivery systems characterised by characteristics of less complexity, and low added value, are dedicated to elementary services, such as policies of the non-life insurance.

It can then be considered that a service that has reached a certain degree of diffusion, and therefore of trivialisation as already known to the public, can also be offered through automated procedures, which do not require a special intervention of the human teller; conversely, for poorly known products, although simple to use, it is important to direct contact with the customer and, therefore, the delivery through a channel that allows interactivity of the relationship, in order to facilitate the promotional activity, as well as demonstrations, of the service.

The increasing diversification of the services offered requires a greater specificity of the delivery systems. The specialisation of the latter can be carried out within the same distribution channel (in particular at the agency), or to give life to relatively dedicated channels.

As is evident, a unique solution is not conceivable, valid for every type of company, since the choices for the optimal combination of the different distribution channels and their organisation derive from assessments on their efficacy and efficiency, and therefore to the potential synergies, or diseconomies, referable to them, not in a theoretical line, but in relation to the specific company reality.

This is conditioned by several factors such as, among other things, the size of the company, the geographic extension of the markets served, and the competitive dynamics related to them, besides, of course, the type of products/services offered and the characteristics of the demand segments.

Moreover, it should not be neglected that the choice of the various types of distribution channels is strongly influenced, by the product policies adopted by the company, the characteristics of the segments served, as well as by inheritance factors. The latter concern, in particular, the existing delivery structure which, clearly, constrains and conditions the strategic choices relating to the rationalisation and restructuring of the entire delivery system, including the enlargement of the range of distribution channels.

The adoption of a delivery system which provides the use of a wider range of distribution channels, endowed with greater specialisation, can allow, in addition to the recovery of efficiency, the achievement of more competitive effectiveness, as the greater specificity of the channel is suitable to satisfy the needs of the target market more appropriately and to optimise, in this manner, the customer relationship (Accenture 2017).

Technological progress enables new delivery systems to be activated, and therefore new distribution channels, in the face of specific combinations of services offered/customer segments served. These channels of distribution are a real innovation of the delivery systems which, in the past, were closely linked to physical support (Stepanek and Roman 2017). As is evident, the automation and outsourcing of part of the production-distributive process and easily standardised services allow the improvement of the operational efficiency and effectiveness of the company's entire delivery system. These, in fact, offer the possibility of a better rationalisation of the entire delivery system and unlock new competitive and marketing opportunities, since they allow you to interact remotely with the user.

The use of technological channels, in fact, in addition to encouraging the structural rationalisation of the distribution network and the reduction of costs to this relative, also allows the improvement of the quality of the services as to the speed and precision of performance, but above all with reference to the possibility to satisfy the needs of the customers increasingly close to the place where they manifest themselves (Normann 2001).

5   INNOVATIVE FORMS OF INSURANCE DISTRIBUTION

In the last 20 years, the traditional intermediaries of the insurance market have to operate in a constantly evolving environment, in which technological innovation and liberalisation have changed the competitive landscape and consequently the distribution mode (Box 4.1).

In recent times, the agency channel has lost importance, at least in part, in the majority of European insurance markets. In France, Italy, and Spain,

---

Box 4.1   Types of Insurance Distribution Channels

- Agents: Intermediaries linked to one or more insurers on behalf of which they distribute products. It is possible to distinguish between the exclusive agents (agents that sell exclusively products of an insurance company) and the multifirm agents (who have agreements with more than one insurer).
- Brokers: Intermediaries independent from insurers, who can normally distribute products of all or most of the insurance companies operating in the market.
- Bancassurance: Distribution of insurance products through bank branches, which includes both purely distributive relationships and cases where there is a common ownership.
- Direct Distribution: Sales are made directly by the insurer without the use of third parties to facilitate the operation. Direct sales can take several forms: in person or at a distance (Internet and other forms of direct sale).
- Price Comparison Websites: Online platforms on which the products of different insurers are listed and compared (especially in terms of prices). From an economic point of view, they can be considered a form of sale through intermediaries because the comparator site acts as a third-party facilitating operations. The aggregators have emerged relatively recently and their effect to date varies significantly depending on the markets.

the use of banks as a distribution channel of life insurance has itself constituted an important trend, whose success is to attribute, in a context of institutional and regulatory liberalisation, to the capillary presence of the bank in the territory, to the availability of detailed database of the customers, to advantages in terms of scale economies. The broker channel has remained stable over time or has lost market shares, especially in the area of non-life insurance products (Accenture 2015).

On the other hand, in the area of elementary products, for which the reasons for choosing the insurance company based mainly on comfort and price, there have been widespread solutions of innovative type, whose production is linked to technology progress in the area of electronics and telecommunications.

It is registered at European level, especially in the UK and the Netherlands markets, a huge expansion of digital channels for the distribution of non-life insurance, whose distance selling takes place via the telephone, websites, and, more recently, emerging aggregators' sites, aimed at the orientation and purchase of non-complex and standardised insurance products such as motor insurance and home insurance (European Commission 2016; Barret 2017).

5.1   Price Comparison Websites

Price comparison websites are Internet-based platforms that offer consumers the possibility to quickly and easily compare the estimates of a given product. Quotes are personalised in terms of the individual's main characteristics and simplify the execution of a purchase based on the search results.

Because of this, a growing number of insurance companies are committing to adopt this channel, implementing a mix of low-cost products specifically dedicated to price aggregators. The approach to using aggregators varies significantly on basis of the geographic location and branch of activity. In 2015, 83% of British insurers were considering launching their own comparison portal over a three-year period, compared to 49% in the rest of Europe and 58% in the USA. Companies operating in the life insurance sector are more open to this approach than their counterparts in the non-life sector (Accenture 2015).

It is possible to classify the business models of price comparison websites based on how revenue is generated (Box 4.2), although these models often show a mix of the different formulas.

Box 4.2   How to Generate Revenue from Price Comparison Websites

- Charge to the Customer: In this case, the costs of the search services provided are charged to the customer. Although theoretically possible, this model is actually the exception instead of the norm. The clear majority of price aggregators do not charge costs directly to the users, but generate revenue in another way.
- Advertising and Sponsored Links: Neither consumers nor suppliers are subject to direct charges in this scenario. The price aggregator, on the other hand, leverages its ability to attract a large number of visitors and charges fees to other companies who wish to place advertisements on the website.
- Charge to Suppliers: This model is by far the most widespread and consists in applying the charges that may take the form of subscription fees, of click-through commissions, of fees conditioned by the positive outcome of a completed operation to the suppliers of the product or services.
- Sale to insurers, or other interested parties, of information collected on consumer preferences.

From the customer's point of view, the main advantage of price aggregators is to reduce the search costs of the best buying solution. These platforms, in fact, provide a quick system to compare quotes, simplifying and reducing the process of comparing the prices offered by the different insurance companies.

The insurers, on the other hand, have the possibility to reach a large number of customers without having to rely on the traditional and more expensive distribution channels. In addition, they can benefit from the massive advertising campaigns put in place by the aggregators, whose charges, however, are partly transferred to the insurers, reducing the actual advantage.

A limit to the use of such portals concerns the presence of possible conflicts of interest with the user if the comparative site exerts an influence on the comparative activity aimed at favouring a company at the expenses of another (listing bias) or uses expressions that tend to sponsor a particular product. In this case, a prejudice is created for the user, resulting from the conditioning of his decision-making autonomy, and a disparity of treatment between companies, so as to alter the balance of fair competition (Baye et al. 2004).

In relation to conflicts arising from trade agreements, it has emerged that the sites mainly compare the products of those companies with which they have entered into partnership agreements and from which they perceive commissions in relation to each contract stipulated (Belleflamme and Peitz 2010). This often results into price comparison websites not offering total market coverage. In addition, some insurers expressly prevent the inclusion of part of their product portfolio. Less complete is the market coverage of the website, the smaller the benefit for users. If consumers think, erroneously, that all existing products have been compared, they can make the purchase under the false conviction that they have purchased the cheapest product available on the market (Caillaud and Jullien 2003; De Cornire and Taylor 2014).

It should be considered that the only comparison criterion used is the price, without considering the various contractual contents, in terms of the coverage offered and the characteristics of the products. However, even in the context of relatively standardised products (such as motor vehicle insurance), there is a certain degree of differentiation. The more the compared product has articulations, the more numerous should be the criteria according to which the comparison is carried out. The comparison based solely on the premium, in fact, does not allow the consumer to assess the actual convenience of the product in relation to its own insurance needs, which may relate to certain essential characteristics of the policy such as the ceilings, the compensations, the exclusions, and the specific forms of settlement of claims (Ronayne 2015).

The good quality of the comparison is often affected by the non-homogeneity of the comparison conducted on products that have different tariff parameters, not envisaged by the customer-compiled estimate module (such as ceilings) or non-uniform presentation standards (e.g. forced combination of ancillary covers, discounts related to credit card payments). The consumer, therefore, is not enabled to make an objective comparison between the products or to obtain a comparison between products tailored on their own real needs (Arnold et al. 2011).

Further limits relate, finally, to the pricing mechanism. In the first place, we refer to the mechanism defined as "dynamic pricing determination," on the basis of which prices may vary upward on the second visit of the site (Gorodnichenko et al. 2015). This, of course, is likely to encourage consumers, if they are aware of this practice, to make hasty purchases.

A second aspect concerns the segmented determination of prices, that is, the mechanism for which the price list on the website may not correspond to the actual purchase price. Insurers may, in fact, impose additional charges (administrative costs, costs for the use of certain methods of payment, etc.), which may not be revealed to consumers until the last step of the online procedure. In order to overcome these information discrepancies, the potential purchaser is obliged to bear higher research costs to determine the actual price of each product (Baye et al. 2006; CMA 2017).

5.2   The Use of Mobile Apps: Trends and Perspectives

Mobile devices have now become a pervasive element in everyday life. These devices have changed the way consumers find information, choose and buy products of all kinds. The use of the computer is increasingly limited, replaced by smartphones and tablets. A survey highlights that in October 2016 the use of the Internet through mobile devices has in fact surpassed the desktop one for the first time all over the world (StatCounter 2016).

The impressive rates of diffusion of smart phones and tablets show how today, and even more in the near future, insurance companies are forced to fully embrace the logic of mobility, implementing technologies suitable for satisfying consumer expectations. Mobile devices, thanks to the development of new applications, become more and more "smart" and all this will increasingly affect the approach to the market, offering new opportunities both to insurers and to customers.

Consumers who take out an insurance policy may request the activation of a reserved area within the company's Internet site to consult in real time the terms of coverage and receive periodic communications. The evolution of this service involves the use of apps for smartphones and tablets that offer additional services and in some cases cost savings compared to traditional delivery solutions (Carney et al. 2015).

Mobile technology changes the way in which the company interact with the customer, creating an additional point of contact, in a sector like the insurance one where interactions with customers are not very frequent.

Besides innovating the traditional mode of underwriting a police, replacing the use of the card with electronic or graphometric signature solutions, mobile apps modify many other aspects of the insurance delivery system, also offering a new distribution channel for consumers that cannot be served by traditional agents.

Through these applications, it is possible, in fact, after the creation of a personal account, to view quotes, to purchase products and services, and to consult in real time one's own insurance position regarding the details of coverage, expiration, claim statements, and so on.

The products, their distribution, and any other activity of delivery will appear very different over a few years, thanks to the capabilities of mobile devices. Sector studies also say that those who do not use such mobile apps would be inclined to do so if the insurance companies were able to better intercept their needs (Carney et al. 2015).

To meet the needs of these consumers, many insurance companies have launched or perfected applications that can assist the customer at every stage of the production/distribution process. In the field of motor insurance, for example, applications are proposed that allow to quickly request the assistance of the roadside rescue, accurately detecting the position of the insured, thanks to the geolocation of the smartphone; in case of necessity, it is possible to find the closest car-body shop, among those affiliated. The management of the claims becomes simpler because the advice of claim can be done via mobile app, attaching photographs of the accident taken with the smartphone; the state of the claim process can be verified at any time, always through the mobile application.

Another opportunity offered by mobile app, in the field of car policies, is connected to the use of the so-called black box, a device that, installed in the car, can be able to compare the parameters of the vehicle and record the data detected while driving. By analysing the data collected from the black box, it is possible to accurately trace the causes of the accident and how it happened. In case of theft, thanks to the use of a GPS detector built into the device, you can be able to trace the vehicle. The use of the black box can control the driving style of the insured and, according to the level of caution recorded, allow discounts on the premium.

The benefits that insurance companies can derive from the implementation of such applications are mainly related to the reduction of information asymmetries and moral hazard behaviours. Gathering information such as timetables, position, and speed, in fact, black boxes help to reconstruct how the accident happened, establish responsibilities, and counter false claims. In addition, the black boxes are also useful for profiling customers, establishing the type of driving and punishing, from a tariff point of view, the unruliest.

In addition to the cars sector, the development of mobile applications also covers other insurance branches, such as healthcare. In fact, especially in the US landscape, there are increasing numbers of companies that use smartphone apps to monitor their customers in this area.

Medical mobile apps monitor, control, and transform data deriving from physiological parameters of the patient. Mobile applications that are becoming more and more widespread in the insurance industry are those that can be connected to so-called wearable devices that allow the monitoring of the physical, cardiac, and sleep activity of the insured. For example, there are applications that measure blood pressure or examine the eyesight, and some applications help the patient in the management of chronic diseases, such as those that calculate the right dose of insulin for diabetics. These technologies assist insurance companies in determining the price of health policies, but also prevent the possibility of fraud in the case of life insurances, allowing to assess the actual risk of the customer.

Among the advantages offered by the wearable/mobile app combination, there is certainly the possibility of devising products that are more customisable and responsive, even in terms of cost, to the risk profile of the insureds, because they are "built" according to the actual and up-to-date data of their health conditions.

However, the use of these technologies raises the big problem of privacy protection. This category of devices presents, in fact, a peculiar level of risk to the privacy of users, because of the suitability of the tools used to detect a set of information including biometric data and sensitive data, such as those concerning the state of health.

In this regard, there are still profound shortcomings, at the legal and regulatory level, as regards the degree of protection of users' privacy by Digital Technologies (GPEN 2016). Some interventions are needed to improve and regulate (1) the ways the data are collected; (2) the methods of storing, using, and erasing the data; and (3) the methods of processing the requests of customers who wish to obtain privacy protection.

6   Conclusions

The changes that have taken place in recent years on the supply side, as well as on the demand side, have meant that a delivery system characterised by traditional networks of an undiversified type could be inadequate to meet the expectations of customers about the type and quality of the services offered. It is essential to adopt strategies of distribution according to the different segments/products that provide for the presence of differentiated distribution channels according to the degree of complexity that characterises the product–customer relationship.

Over the last years, insurance companies have been investing in their processes to make their services more available to customers. This has been driven not only by customer expectations and a desire to increase their convenience and control but also by the need to reduce distribution costs.

However, a deeper transformation is inevitable, since a dizzying succession of technological and other innovations are challenging the traditional insurance business model.

Over the next five years, sensors, the cloud, connected smart devices, and real-time analytics will combine to deliver a new layer of connected intelligence that will revolutionise the ability of insurers to offer interesting and increasingly indispensable digital services to consumers. Insurers are moving steadily towards a digitally enabled omnichannel distribution model. Every stage of the sales process is affected, from discovery of information through to advice and purchase.

In this regard, problems may arise in relation to the possible overlap of distribution channels which are accompanied by inconsistencies of different natures, not least those relating to a possible disorientation of some customers. These, in fact, if not properly informed, can show some discomfort in the face of an innovation, which becomes too accentuated by the ways of service distribution.

Moreover, it should be remembered that there are user segments that are particularly prone to the use of self-service structures and others, on the contrary, that prefer the use of more traditional channels also for the enjoyment of elementary services. As is obvious, these preferences are a strong conditioning in the choice of different options at the operational level.

Therefore, most insurers expect physical channels to endure in the digital age. They envisage a vital role for agents and brokers, albeit one that is markedly different than that which most of them fulfil today.

Finally, the effects in terms of image and reputation that the introduction of such technological solutions tends to produce towards the public should not be neglected, allowing it to bring the idea of efficiency and considerable capacity of renewal and continuous adaptation to the evolution of the market and the needs of the customers. In this sense, the objective pursued by the introduction of new solutions can be the consolidation, if not even in the increase, of the market share.

The decisions the insurers make today will determine not only the kind of customer experiences they will provide to remain competitive but in fact the kind of businesses they will be in the years to come.

REFERENCES

Accenture. (2015). Digital Insurance: Reimaging Insurance Distribution.

Accenture. (2017). *The Future of Insurance Distribution.* New Models for a Digital Customer.

Allmendinger, G., & Lombreglia, R. (2005). Four Strategies for the Age of Smart Services. *Harvard Business Review, 83*(10), 131–145.

Arnold, M. A., Li, C., Saliba, C., & Zhang, L. (2011). Asymmetric Market Shares, Advertising and Pricing: Equilibrium with an Information Gatekeeper. *Journal of Industrial Economics, 59,* 63–84.

Barret, C. (2017, September 29). Price Comparison Websites Are Far from Epic. *Financial Times.*

Baye, M. R., Morgn, J., & Scholten, P. (2004). Price Dispersion in the Small and in the Large: Evidence from an Internet Price Comparison Site. *Journal of Industrial Economics, 52,* 463–496.

Baye, M. R., Morgn, J., & Scholten, P. (2006). Information, Search, and Price Dispersion. In T. Hendershott (Ed.), *Handbook on Economics and Information Systems.* Amsterdam: Elsevier.

Belleflamme, P., & Peitz, M. (2010). Platform Competition and Seller Investment Incentives. *European Economic Review, 54,* 1059–1076.

Black, N. J., Lockett, A., Ennew, C., Winklhofer, H., & McKechnie, S. (2002). Modelling Consumer Choice of Distribution Channels: An Illustration from Financial Services. *International Journal of Bank Marketing, 20*(4), 161–173.

Brady, T., Davies, A., & Gann, D. M. (2005). Creating Value by Delivering Integrated Solutions. *International Journal of Project Management, 23*(5), 360–365.

Brown, J. R., & Goolsbee, A. (2012). Does the Internet Make Markets More Competitive? Evidence from the Life Insurance Industry. *Journal of Political Economy, 110,* 481–507.

Caillaud, B., & Jullien, B. (2003). Chicken & Egg: Competition Among Intermediation Service Providers. *RAND Journal of Economics, 34,* 309–328.

Carney, E., Ensor, B., Ask, J. A., Berdak, O., Causey, A., & Blumstein, A. (2015, February). The Future of Insurance Is Mobile. Forrester.

CMA—Competition and Markets Authority. (2017, September). *Digital Comparison Tools Market Study.* Final Report.

Coelho, F., & Easingwood, C. (2005). Determinants of Multiple Channel Choice in Financial Services: An Environmental Uncertainty Model. *Journal of Services Marketing, 19*(4), 199–211.

Coelho, F., & Easingwood, C. (2008). An Exploratory Study into the Drivers of Channel Change. *European Journal of Marketing, 42*(9/10), 1005–1022.

Coelho, F., Easingwood, C., & Coelho, A. (2003). Exploratory Evidence of Channel Performance in Single vs Multiple Channel Strategies. *International Journal of Retail & Distribution Management, 31*(11), 561–573.

De Cornire A., & Taylor, G. (2014). *Quality Provision in the Presence of a Biased Intermediary*. Working Paper.

Dumm, R. E., & Hoyt, R. E. (2003). Insurance Distribution Channels: Markets in Transition. *Journal of Insurance Regulation, 22*(1), 27–47.

Easingwood, C., & Storey, C. (2006). The Value of Multi-Channel Distribution Systems in the Financial Services Sector. *The Service Industries Journal, 16,* 223–241.

Eiglier, P., & Langeard, E. (2003). *Il marketing strategico nei servizi*. Milano: McGraw Hill.

European Commission. (2016). European Economic Forecast, Winter.

Fürst, A., Leimbach, M., & Prigge, J. K. (2017). Organizational Multichannel Differentiation: An Analysis of Its Impact on Channel Relationships and Company Sales Success. *Journal of Marketing, 81*(1), 59–82.

Gorodnichenko, Y., Sheremirov, V., & Talavera, O. (2015). *Price Setting in Online Markets: Does IT Click?* Berkeley Working Paper, University of California.

GPEN—Global Privacy Enforcement Network. (2016). Annual Report.

Heinhuis, D., & de Vries, E. J. (2009). Modelling Customer Behaviour in Multi-Channel Service Distribution. *Enterprise Applications and Services in the Finance Industry, 35,* 47–63.

Kabadayi, S., Komarova, L. Y., & Carnevale, M. (2017). Customer Value Creation in Multichannel Systems. The Interactive Effect of Integration Quality and Multichannel Complexity. *Journal of Creating Value, 3*(1), 1–18.

Klotzki, U., Gatzert, N., & Muenstermann, B. (2017). The Cost of Life Distribution in Europe. *The Geneva Papers on Risk and Insurance, 42*(2), 296–322.

Kotler, P. (2001). *A Framework for Marketing Management*. Upper Saddle River: Prentice Hall.

Kotler, P., & Pfoertsch, W. (2011). *Ingredient Branding: Making the Invisible Visible*. New York: Springer.

Nightingale, P. (2003). Innovation in Financial Services Infrastructure. *The International Handbook on Innovation, 3,* 529–547.

Normann, R. (2001). *Service Management: Strategy and Leadership in Service Business* (3rd ed.). New York: Wiley & Sons.

Pires, C. P., Sarkar, S., & Carvalho, L. (2008). Innovation in Services—How Different from Manufacturing? *The Service Industries Journal, 28*(10), 1339–1356.

Ronayne, D. (2015, October). Price Comparison Websites. Warwick Economics Research Papers Series, 1056.

StatCounter. (2016, November). Mobile and Tablet Now Gets More Usage than Desktop.

Stepanek, L., & Roman, P. (2017). Urban Insurance Industry Ideas of Second Millenium. *Ecoforum, 6*(1), 10.

Su Chen, M., & Chang, P. L. (2010). Distribution Channel Strategy and Efficiency Performance of the Life Insurance Industry in Taiwan. *Journal of Financial Services Marketing, 15*(1), 62–75.

Su Chen, M., & Lai, G. C. (2010). Distribution Systems, Loyalty and Performance. *International Journal of Retail & Distribution Management, 38*(9), 698–718.

Thornton, J., & White, L. (2001). Customer Orientations and Usage of Financial Distribution Channels. *Journal of Services Marketing, 15*(3), 168–185.

Thornton, J., & White, L. (2009). Financial Distribution Channels: Technology Versus Tradition. *Journal of Professional Services Marketing, 21*(2), 59–73.

Trigo-Gamarra, L., & Growitsch, C. (2010). Comparing Single and Multichannel Distribution Strategies in the German Life Insurance Market: An Analysis of Cost and Profit Efficiency. *Schmalenbach Business Review, 62*(4), 401–417.

CHAPTER 5

InsurTech and Customer Relationship

Abstract The chapter aims to analyse the technological evolution impact on customer relationship types and strategies in the insurance field. The loss of the interpersonal relationship, which is the direct consequence of the adoption of digital distributive forms, calls for a correct setting of the market relationship. Indeed, the digitalisation of the offer entails a greater involvement of the customer in the service delivery process, who can judge this circumstance more or less favourably. The chapter highlights how this requires a personalised approach to the market, on the basis of a logic of integrated communication and customised marketing. As to the latter, the various direct marketing strategies and techniques are aimed at the customer loyalty through the interactive use of the several media which are at the company's disposal.

Keywords Digital customer management • Direct marketing in insurance • Insurance customer relationship • Customer loyalty

1 INTRODUCTION

The relationship between an insurer and its policyholders is typically infrequent, fleeting, and of a financial nature. Because of this historical interaction model, for a long time insurers didn't need to adopt cutting-edge technologies. In fact, at a time when many other businesses have migrated from legacy systems to mobile and cloud solutions, insurers are still some

© The Author(s) 2018   75
A. Cappiello, *Technology and the Insurance Industry*,
https://doi.org/10.1007/978-3-319-74712-5_5

of the largest users of mainframe technology. But this is changing. Insurance industry, like many other industries, is facing sweeping changes driven by a confluence of business and technology forces fueled by innovation. A new wave of insight, or interactions, and value is bringing insurers and their personal and commercial customers closer together. Insurance companies understand that they need to become more customer-focused, easier to do business with, and more responsive to customer needs and expectations.

Evolving consumer behaviour is threatening traditional marketing levers, such as television advertising; hence it is necessary to adopt personalised strategies and techniques to interact with current and potential clients and improve the relationship between company, channels, and customers.

2   TECHNOLOGICAL INNOVATION AND EVOLUTION OF CUSTOMER RELATIONSHIP

The supply of services provided with automated procedures, which are able to transfer the distribution as close as possible to the potential customer, has the logical consequence of a substantial transformation of the relationship with the latter.

The use of digitised channels, in fact, modifies the elements that compose the traditional delivery system, and the relationships that come to be established among them. As is evident, the overcoming of the direct relationship with the sales staff implies a stronger intervention and, therefore, a greater involvement of the customer in the process of delivering many services (Badoc 1986; Boulding et al. 1993; Eiglier and Langeard 2003).

The diffusion of new technologies, which, as said, affects the behaviour requested from customers, and the degree of their participation in the process of delivery of the service, can be judged by these more or less favourably.

There are, in fact, market segments which are satisfied by the use of automated procedures which, without the need for any intervention by the human teller, allows to benefit from a service available throughout the day and directly to your home (Cummins and Venard 2007).

On the contrary, there are market segments traditionally wary of innovations, which are afraid of technological applications because of not being able to use them or because they consider them unsafe and easily accessible to strangers. This range of users tends to consider the introduction and diffusion of channels with high technological content as a factor that worsens the level of service quality because of the greater complexity of the behaviour required, compared to the traditional distribution forms (Dumm and Hoyt 2003; Thornton and White 2001).

However, it is certain that, in time, the users will tend to become more familiar with self-service procedures; this will determine the transition to a more autonomous and conscious attitude by the customer who, called to participate actively in the process of delivery of the service, will be increasingly attentive to the aspects of quality, efficiency, and cost.

It should, however, be stressed that, on the one hand, IT significantly contributes to improving the quality of service by reducing waiting times, expanding service production times, reducing the number of accounting errors, and so on; on the other hand, it presents the disadvantage of making the insurer–customer relationship more impersonal.

The new technologies eliminate, on many occasions, the direct relationship between the insurer and the customer and therefore the possibility of an interactive communication between the two parties. This becomes more relevant in relation to the intangible nature of the insurance service, which is only easily assessed in its qualitative aspects at the time of use. The constituent elements of the service delivery system therefore have a significant impact on the quality, as perceived by the customer; this can be understood as the result of the combination of different elements such as the ease of access and convenience of enjoyment, the speed of time and reduction of material errors, the transparency and the cost-effectiveness of the relationship, and the professionalism and courtesy of the intermediating employee who manufactures and at the same time sells the service (Donnelly et al. 1985; Normann 2001; Kotler 2001; De Ruyter et al. 1998).

As is evident, technological innovation offers the opportunity to improve many of the aspects considered; however, this tends to cancel the benefits of personal contact between the customer and the insurer, since the latter does not appear to be part of the delivery system if the service is distributed using digitised techniques. This also reduces the possibility of proposing differentiated solutions in relation to demand requirements, or to conduct cross-selling targeted actions.

It should also be noted that the depersonalisation of the relationship with the customer can tend to reduce the customer's loyalty to the institution, as quality judgements are increasingly expressed on the basis of considerations of technical-economical nature, and increasingly less on the basis of emotional factors (Reicheld 1996; Schwarz et al. 2014).

Another aspect to consider is the intensification of the competition between insurance companies due, among other things, to the most frequent overlap of areas of expertise made possible by the growing spread of computer technologies.

In this respect, the functions of promotion and sale, distinguished from the mere technical distribution of the service, become essential to fruitfully manage the relationship with customers and to overcome, in this way, the depersonalisation due to the introduction of new technologies (De Ruyter et al. 1998; Patterson and Ward 2000).

It is to be observed, in this regard, how the technological variable should be managed by the company not only in a purely productive key (to streamline the procedures and reduce operating costs), but also in a marketing perspective to improve the corporate image. In fact, the customer's judgment on the individual service and the one on the company delivering it tend to coincide.

The offer of certain innovative services, and distributed by means of non-traditional channels, also contributes to the diffusion of a modern, efficient, and competent image to the public. The promotional action, or more generally marketing, must therefore be set in this sense: to obtain a more massive dissemination of computerised services to the insureds, without losing sight of the opportunity to establish with the same an interactive and loyal relationship.

The loss of direct contact with the customer is undoubtedly a negative factor that cannot be neglected in the field of insurance, nor in the programming of different marketing policies. It is therefore necessary to balance the need of the personalised relationship with the customers with the benefits offered by the automation; this objective can be achieved by seeking a marked standardisation of the elementary services to which a greater specialisation of the company–client relationship needs to be associated.

In fact, the increase and differentiation of customers' needs indicate that a market orientation that does not take into account the specific characteristics of the same is now outdated.

In this regard, it is appropriate to adopt a personalised approach to the market, according to the logic of customised marketing. This approach, without considering the market an undiversified whole, provides for the identification and development of particular segments or micro-segments in order to achieve, in certain cases, the establishment of relations on an individual basis. Through a more precise identification of different needs and different requirements, it is possible to obtain a greater personalisation in the delivery of services and in communication with the reference market (Nash 1995).

The above considerations underline the need to recover a direct contact, more qualified and personalised, with the customer through channels to facilitate the conduct of an interactive communicational activity. First, we refer to the agency network and the activity of the sales staff destined to contact with the public, as well as to the different strategies and techniques of direct marketing, aimed at customer loyalty through the interactive use of the different media available to the company.

3   Customer Loyalty and Direct Marketing

Currently, the relationship with the market is still characterised, in most cases, by the use of traditional distribution channels (agency network) and by the almost totally unidirectional communication; in this context, the interactive company–client–company relationship can only develop through the intermediating employees at the point of sale, or by contacting the dealers (personal selling).

On the other hand, given the increasing automation of procedures, the innovative distribution channels, if not used interactively, contribute to a progressive depersonalisation of the insurer–customer relationship. Moreover, it should not be neglected that certain circumstances, including the increasing competition between insurance institutions, as well as the economic and cultural evolution of the customer, contribute to increasing the mobility of the latter by reducing the intensity of the insurance bond.

It is therefore necessary to seek continual improvement in the communicational approach with the reference market in order to try to renew and innovate, where possible, the relationship with the customer in view of a more complex fulfilment of expectations of the latter (Brioschi 1990).

While, on the one hand, the tendency to use automated channels is growing for the simplest services, on the other hand, especially for more complex services, the demand for a greater personalisation of the relationship and communication with the company is increasing.

The clients, because of the growing economic culture, emerge from the mass to position themselves on the market in an individualised way. A higher personalisation of services and communication with the market can be achieved effectively through the introduction, also in the field of insurance, of strategies and techniques of direct marketing, in the light of experiences already gained in other areas of activity (Peng and Mercer 1998; Hänninen and Karjaluoto 2017).

Direct marketing can be defined as a strategic process aimed at increasing the loyalty of the customer (or segment) by creating a personalised and especially biunique relationship with the consumer (Rowley 2005). To this end, direct marketing uses interactive tools that can address a targeted audience and push it to an action that generates measurable answers. In short, direct marketing aims to activate a two-way communication channel between the company and its market through the use of interactive means and through the management of a specific database (Baier 1985; David Shepard Ass. 1995; Nash 1995; Roman 1995; Brian and Housden 2017).

Therefore, the three elements characterising direct marketing are outlined here: the interactivity of the communication channel, the measurability of results, and the personalisation of the relationship with the consumer (Roberts 1989).

With regard to the latter aspect, it is useful to underline how direct marketing strategies reject the so-called mass approach to the market, understood as an undiversified ensemble, but as a sum of individuals, each of whom is distinguished by their own individuality and personal needs.

As is evident, the personalisation of the insurer–client relationship, made possible by the use of a channel with a double sense of communication, allows a better loyalty of the insurance user who can turn, more likely, into a company's long-lasting customer.

In fact, the process of "clientelisation" involves the search, identification, and customer loyalty over time, in order to achieve an auto consolidated growth of the market. From what is said, it is obvious how direct marketing implies a concrete orientation to the customers, allowing to establish with these a more immediate contact and to provide, at the same time, concrete information about the product delivered (Oliver 1999; Kroll 2016).

In summary, therefore, direct marketing contributes to improving the company–client relationship, to spread a positive image of the company, and to disseminate a wider knowledge of the different services offered; at the same time, by means of a continuous feedback process, it is possible to correct and adequately plan the different market policies.

3.1   Direct Marketing Tools

The increase of consumer's loyalty, according to the logic of direct marketing, can be achieved using interactive media, used in a synergistic way and integrated with the other communication tools available to the company. The techniques of direct marketing therefore assume equal respect to the means of the traditional communication mix, thus going to form a global communication mix of the company (Delanote et al. 2013).

The media available to the company to implement a direct marketing strategy can be divided into mass media, however used interactively, and properly interactive media (Holtz 1986).

Newspapers, periodicals, radio, television, billboard, the cinema, and so on belong to the first group.

If, as is often the case with insurance products, we do not have sufficiently large mailing lists regarding a specific product, it is advisable to use the press or radio and television (especially in the first phase of market screening). These means of communication achieve a good level of effectiveness in the context of direct marketing strategies if it is possible to activate a response process by the public.

The mass media used interactively have the advantage of reaching a large part of the market at a much lower cost per contact than that of the properly interactive means; it is also possible to reach specific geographic areas—for a company, for example, the own area of activity—or particular demographic categories. On the other hand, it should be noted that these means have habitually lower levels of affinity, understanding the latter as the logical connection between the supply and the potential customer to which it is addressed.

The media that are properly interactive include, in addition to direct mailings and telemarketing, all the electronic media that offer the possibility, to the one who receives the message, to activate a response behaviour. These means are therefore marked by the possibility of a dialogue, as they reproduce the human characteristic of communication, namely the biuniqueness of the flow of information.

When it is not possible to establish a personal relationship for technical or economic reasons, we can switch to the technological approach of communication that involves relatively minor costs and can cover larger areas.

Direct mailing consists of sending booklets, flyers, printed advertisements, and several other types of texts, all accompanied by a letter of presentation, which is in any case indispensable.

The fundamental objective is to arouse an action, a behaviour on the part of those who receive the message. Such behaviour may be a purchase, or only the request for further information or for the visit of a seller at home. In the context of the promotion and sale of insurance services, a direct mailing is plausible to stimulate an action of the second type, aimed therefore at communicating the business of the company, to inform the user and to stimulate their interest for a further contact, digitised or personal.

It is not possible to exclude an a priori use of this instrument as an alternative channel of distribution of banking services, perhaps of a simpler type, and in compliance with the rules in force in this field.

It is apparent from the foregoing that the advantages of direct mailing are attributable to the absence of geographical limitations and the possibility of selecting the market and of establishing direct contact with the recipient; moreover, through the mailing you can provide a series of information about the product service, as well as test easily all the variables of the offer, with the opportunity to obtain a rigorous analysis and interpretation of the results.

Implementation costs—including the creative approach, the technical processing, and printing and shipping costs—can be considered an overall acceptable investment as it is circumscribed to a well-identified target, which the communication is directed to, without dispersion on sterile segments.

It should be noted that direct mailing clashes with two types of obstacles: the first one is linked to the problems of technical-operational order of the Postal Service, and the second one relates to the distracted attitude of the public towards a communication of that kind. It should be noted, in this respect, that the insurance company still has a privileged position compared with other types of companies because, at least for the current customers, the mailing is often shipped together with the statement, managing to arouse caution in the player. Also with regard to potential customers, it can be considered that messages sent have good chances to stimulate the recipient's attention as the public shows growing attention to the news of insurance/financial order and as regards the new products offered by the insurance in the social security sector.

It is necessary, however, to integrate the mailing with other instruments (such as telemarketing) as this instrument must be considered as part of an entire communication system; it is therefore necessary to reinforce the action of the mailing with media support that will create, from the first moment, a strongly interactive contact with the customer.

Telemarketing concerns the activation of an inbound (receiving telephone calls) or outbound (making telephone calls) telephone service. With the first, making available to the public a dedicated number, the company responds to requests for information, provides a first consulting service, or can schedule a subsequent personal contact. Finally, it is not to be excluded that the user, through this channel, can directly request the purchase of a particular service. This type of telemarketing can be considered the final phase of a communicational campaign conducted through the mailing or the traditional media (e.g. the press), the effectiveness of which can be easily tested, thanks to the high potential of information feedback that this channel owns.

With the second method, more selective, the company becomes an active part, taking the initiative to communicate with the customer. It is, in this case, to conduct an interview, to provide some information, or even to stimulate a purchasing behaviour. It follows that the use of telemarketing, as already noted for direct mailing, may also constitute a strategy for diversifying the distribution channels of the insurance company.

As is evident, the use of this channel allows to establish a strongly interactive type of communication, second only to the personal relationship, and to obtain readily controllable and measurable results.

Because, clearly, the cost per contact is high, the use of telemarketing can be justified due to its considerable selectivity. It is necessary, in this regard, to use the telephone according to a systematic scheme of objective definition, target setting, message building, and analysis of results. In fact, it would not be useful and would also be extremely expensive to contact all customers of the branch without an adequate planning and organisation.

In summary, the main advantages of telemarketing can be traced back to the immediacy and a strong personalisation of the contact, the ease of response, and the flexibility and adaptability of communication; the interlocutor can in fact adapt the conversation to the different reactions of the client.

A shortcoming to the effectiveness of telemarketing is that the call can be considered by the customer as a violation of their privacy. In this sense, it is advisable to use this type of contact in the case of existing relationships.

Finally, it should be noted that although, on the one hand, it is certain that the anonymous telephone operator trained to be polite, up-to-date, and fast can replace the agent in an excellent way, on the other hand, we should not overlook the fact that the use of telemarketing implies the loss of that strictly personal contact which is often a decisive element in the various marketing policies. It can therefore be assumed that telemarketing techniques should not be used indiscriminately for any type of service, especially when a face-to-face contact is certainly more profitable which, in any case, must be constantly urged.

Other media characterised by communication interactivity can be used in direct marketing techniques. We are hereby referring to tools that computer technology continually puts on the market, from social networks (Facebook, LinkedIn, etc.) to the different types of mobile apps and so on, through which the banking company can activate a direct channel of communication with its own audience.

It is evident that there is a close correlation between interactivity and the cost of the different media; this means that the cost per contact rises as the communication becomes more effective on the response behaviour. If we wanted to exemplify the assumption by using a diagram, the maximum values would be relative to the personal sale, while the lower ones would concern the radio and television communication.

Also in relation to that aspect, it is necessary to fix an appropriate communication mix. In the market screening phase, for example, it may be sufficient to have a communication carried out by means of press or radio and television (used, of course, always with the possibility of response) which have a lower cost (per contact); in an early phase it is important, in fact, to be able to reach a market coverage as wide as possible (Nash 1995).

Later, when a relationship is established with the customer, it is advisable to start using a range of more interactive media, even if more expensive. At this stage, the relationship becomes productive and there is security of a return of the investment; moreover, only through the biuniqueness of the communication it will be possible to carry out a customer loyalty action (Bloemer et al. 2009).

3.2   Direct Marketing Opportunities and Limitations

The consideration outlined above confirms that direct marketing has an enormous potential, and certain limitations, when applied to insurance management.

The primary advantage that is connected to direct marketing strategies is the possibility of establishing direct, interactive contact with the market, to which the benefits deriving from the characteristics of concentration and customising of the message are also connected (Su Chen and Lai 2010).

It is clear, in fact, how direct marketing allows to direct (concentrate) communication to the most likely users (thanks to an accurate segmentation of the interested market) and also to personalise the same according to the expectations of a single customer.

In particular, besides allowing the achievement of considerable cost savings with regard to the promotion activity, it also offers the opportunity to obtain the measurement of the return effects related to the same. Thanks to the use of different direct marketing techniques, it is possible to conduct tests on advertising campaigns, for example, by making exploratory mail forwards, or by diversifying ads. In this way, it is possible to measure the results and to make statistical projections that present a good level of trust. Clearly the latter, of a strictly quantitative nature, must also be accompanied by qualitative research, which studies the "why" of such behaviour, or of a given reaction of the public. Only to start, it is possible to conduct an analytical study about the needs and requirements of the public, to try to satisfy and even to anticipate those expectations with timeliness.

The measurability characteristic of the results brings, as a logical consequence, the flexibility of the different direct marketing policies, as well as of the different market strategies.

It is known that the measurement of the results of traditional advertising campaigns (i.e. through the mass media) is a problem of not easy and unique solution; on the contrary, in direct marketing, the ability to carry out limited tests, and to quantify the results in a relatively short time, permits a timely adaptation of the promotional action, and more generally of every aspect of management, to market variables, with better performance in terms of efficiency and competitive effectiveness.

In expressing a judgement of convenience on direct marketing, we must not neglect the characteristic of immediacy that characterises this type of communication.

Conventional advertising is aimed essentially at establishing the awareness of the product and encouraging a positive attitude (to give birth to desire), which can be nullified during the distribution phase. In the specific case of the insurance company, a crowdy agency, a too expeditious or poorly informed (and trained) employee, and an inaccessible site could definitely demotivate the customer.

Direct marketing messages, on the contrary, whatever the means used, are aimed at soliciting immediate behaviour, whether this be a purchase or even a simple request for further contact, be it telematic or personal. In this way, it is possible to overcome the inertia of the consumer (Brasini and Tassinari 2014).

Thanks to direct marketing, the cost of wasting time to go to the agent to ask for information or to purchase a particular service is cancelled. From your home, without having to deal with traffic, search for a parking space, and stand in the queue at the point of sale, the user can obtain the requested information and, thanks to technological progress, enjoy an ever-wider range of services.

In fact, direct marketing is a factor of innovation and differentiation of the distribution network, since the operational instruments that relate to this can be considered, in some circumstances and on given conditions, real distribution channels of service. The simple communication—activated both with the most traditional interactive media, and with the most advanced ones—can be immediately followed by the moment of the real delivery of the service. In this way, the communication process is included into the most complex production and distribution of the product (Kotler and Pfoertsch 2011).

Where the use of technology implies, as is natural, the loss of a personal contact with the customer, direct marketing offers the opportunity, as we have already mentioned, to recover an interactive and loyal relationship, which can replace, in part, the one activated at the physical point of sale.

It is clear in this regard that direct marketing contributes to differentiation (and innovation) not only of the distribution process, but also of the product-service offered. Given, in fact, the nature of the intangibility of the banking product, which significantly limits the possibilities of differentiation, it is intuitive that direct marketing should be considered an effective tool of differentiation of the service itself, thanks to the use of direct and personalised communication (Normann 2001).

In addition, direct marketing can contribute to the targeted and widespread dissemination of information at the market segments to be achieved. This aspect is particularly important in terms of insurance, due to the peculiarities of the offer and the limited spread of economic and insurance culture in particular, which still characterises certain segments of the market and certain geographic areas.

It is known that the insurance product, not presenting tangible requirements, needs a communication that is able to establish a direct contact between the deliverer and the user of the service. In this way, it is possible to carry out a more massive action of information and advice regarding the different technical ways of use of the service itself, as well as the economic advantages of the various possible alternatives (Rust and Zahorik 1993).

However, such information is all the more valuable the more the customer does not know certain services or feels an unmotivated mistrust towards them.

Finally, it should be noted that with direct marketing it is also possible to carry out an effective cross-selling action. In fact, once in possession of certain lists of customers satisfied of the relationship with the company, the latter, after an accurate segmentation according to parameters deemed to be more convenient, is able to propose and therefore to sell complementary, alternative, or simply distributed services to the same customers through the insurance channel. It is certainly easier to develop an existing link than to create a new one.

It cannot be neglected, however, that a too massive use of direct marketing techniques may result in a further reduction of contact with the customer, in cases where the latter would have preferred meeting an agent personally (Wang et al. 2017). It is therefore necessary to have a suitable programming of direct marketing communication which must be reserved for contacts that could not be implemented otherwise, without sacrificing the personal insurer–client relationship whenever this is necessary or requested by the customer.

Another problem that needs to be carefully assessed is the risk of seeing, in the customer's mind, the company's direct marketing activity associated with promotional selling campaigns carried out, and sometimes incorrectly, by companies belonging to different market sectors (e.g. commercial companies) (Mela et al. 1997; Bird 2007).

Due to this, it is possible to find a certain degree of inurement to this kind of marketing. This disinterest can only be overcome with a direct marketing strategy whose main requirements are transparency and professionalism, and which indicates, at least at the initial stage of the market survey, an activity of information and completion of other communication strategies.

It should also be stressed that the adoption of a concrete direct marketing strategy by the insurance companies requires, consequently, the organisation of human resources, not only those that are required to carry out the activity in question.

It does not seem obvious to remember, once again, that direct marketing cannot and should not be identified only with the activation of the toll-free number or by sending hundreds of customer messages, since these do not represent the only aspects, albeit important, of that strategy. The absence of real coordination, at all hierarchical levels, may become the cause of failure, in fact, even of a campaign of sure success.

Another factor of development, which can represent a limiting aspect, is given by the availability of technological equipment and their diffusion. The technological resource, which will become an increasingly propulsive factor for the direct marketing techniques in the future (which will also be transformed by it), can be analysed as an internal variable in the company or as external variable.

It is necessary to acquire highly flexible and specific technologies, which enable, on the one hand, the collection, interpretation, and use of information for the elaboration of personalised promotional campaigns and which promote, on the other hand, the streamlining and improvement of the communication networks with the market.

As far as the technology intended as an external variable is concerned, we are hereby referring to the diffusion of technology, particularly IT, to the public.

From the identification of the main advantages and limits associated with the direct marketing strategy, it is evident how the latter, if introduced in close correlation with other marketing strategies, can produce good results.

In practice, the insurance company, with the use of personalised, interactive, and loyal communication, can achieve a twofold series of objectives: on the one side, the increase in demand by already acquired customers, and, on the other side, the promotion of the first contact with potential customers in order to transform it into effective customers (Javalgi and Moberg 1997; Iacobucci 1999; Butcher et al. 2001).

If appropriately included in the planning of the different distribution modalities, direct marketing can also contribute to the rationalisation of the production and delivery of the service processes.

Obviously, direct marketing techniques cannot be used for all insurance services in an indistinct way. Certainly, direct marketing can find a good application regarding the promotion and sale of products services of non-complex type; for products that have a higher level of complexity, direct marketing can be useful as a marketing tool supporting other actions, and to increase the awareness of the potential customer through direct information, or by giving them the possibility to request information, at least at first approximation. At a later time it seems essential, for this second type of services, the constant and specialised intervention of the intermediating employees to treat the delivery of the service directly, through a thorough consultancy activity.

It therefore follows that for simple products, thanks also to the progress of IT, it is possible to use direct marketing to also develop a sales action through the use of different technological interactive media, besides a purely promotional activity.

On the other hand, for high-quality products, direct marketing can be used primarily to inform customers, rather than to prompt the user to search for further contact, thus leaving other operating areas the real selling function; this is due to the complexity of the product in question.

Only with the transition from a traditional marketing approach to one of direct marketing it is possible to recover, in the relationship with the customer, a good level of interrelations, too often neglected and sometimes made impossible by the more and more widespread use of IT solutions, which permits to deliver an ever-wider range of services with dematerialised modes.

However, thanks to the use of these technologies, direct marketing is able to cope with the need for individuality of the client and to establish with them a long-lasting relationship, with consequent undeniable advantages also in economic terms.

4   CUSTOMER RELATIONSHIP MANAGEMENT AND SOCIAL MEDIA STRATEGY

Social media have now become a consolidated mass phenomenon that contributes to change the lifestyles of people; the communicational, personal, and professional environment; and the ways through which public opinion is created (Viale and Souche 2012).

An increasing number of people use social media in order to acquire useful information to form their own purchasing decisions, also thanks to the creation of forums of opinions in which the participants exchange recommendations and experiences.

Because of their massive diffusion, the new social interaction tools are clearly penetrating even into the insurance sector. Therefore, it is necessary to understand in time the importance of social media in the construction of effective communication strategies (Capgemini 2014).

Social media constitute a unique challenge for insurance companies and offer opportunities for the development and creation of a lasting and interactive consumers' relationship, allowing them to establish a highly personalised contact with some people or with the community. These platforms are not only a customer service tool: they also boast a huge potential for service promotion and also produce an important amount of data about user behaviour, that is to say reusable information in order to improve company delivery system.

Among the reasons that lead the companies to decide to enter the social media, the first one is the desire to develop the brand value, followed by the willingness to retain customers, and the search for a contact with potential customers (Insurance Europe 2017; Goodman 2014).

There are significant differences between traditional companies and those that are entirely based on online platforms. The first ones adopt more institutional social communication strategies, while the latter demonstrate greater familiarity and creativity in the use of digital instruments.

The use of social media permits the increase of the company brand value by seeking a new paradigm of communication and relationship with the company.

By monitoring the content on the social media, it is possible to identify the "brand sentiment" of users and, if negative statements appear on the network, take proactive measures to promptly address the emerging issues which, if ignored, could lead to considerable reputational damage. The perception of the public towards the insurance sector is not always positive, and companies often have to promote campaigns focused more on creating trust than on the characteristics of the service offered. This is why social networks play a decisive role for this sector, where disgruntled customers can be intercepted and reassured, and it is possible to build a lasting trust relationship with them.

According to the World Insurance Report, the main problem is the dissatisfaction of the technology-expert youngsters, who have become more demanding. The digital consumers who express themselves and move through social media are deeply different from the clients to which companies and agents are accustomed. Thanks to the pervasiveness and dissemination of online communication platforms, the new consumer becomes more autonomous and more critical of the insurance offer. More and more informed and documented on the countless areas surrounding the purchasing choice, the social customers tend to evaluate their choices more depending on their mood, than on the basis of a loyalty to the company that is increasingly being worn. Compared to routine behaviours that characterised the customer in the past years—traditionally linked to the brand by a relationship of hereditary kind—the new consumers of the digital era finalise their purchases only after a careful selection of offers and promotions on the market, with particular attention to the concerned distribution channel. Customers are therefore living a radical transformation becoming much more competent: they call for the composition of the products and are often aware of the technical aspects related to the individual policies, thanks to the many sources through which they can accurately compare products. The new customer mainly trusts his network of knowledge, his personal social network, in which the classical theories of marketing are no longer valid.

Understanding their strengths and weaknesses is an activity that is now essential for insurance companies, if they want to avoid a reputation problem that could damage what has been built over time. In fact, the information posted about an insurance company, agency, or intermediary can be read and commented by anyone who has access to the Internet. Even a single user on the network can develop a discussion, gathering around them other subjects with whom to vent their dissatisfaction towards a particular company or brand.

The customer care service is traditionally based on the one-to-one relationship between the company and the client, but thanks to the social networks, customer service can also be based on a many-to-many logic and become another moment of sharing and participation. The participatory logic causes the resolution of a problem to also occur by exchanging information between users. By leaving this mutual assistance to the consumers themselves, the company can avoid repeatedly responding to the same questions: it is precisely in solving the simplest problems that consumers themselves can provide answers and, in some cases, even more effectively than the customer service, thanks to their personal experience.

A strategy based on using social media can help businesses to reduce call centre costs by redirecting service efforts and resources to more critical business areas.

A further advantage offered by social media to insurance companies is the strengthening of customer relations. In fact, the use of these platforms contributes to influence the customers' perception of the company, also thanks to the word of mouth of the customers who have had positive experiences and who would recommend the brand to others.

Customers are increasingly keen to be protagonists; the listening capacity becomes, therefore, fundamental to understand the expectations and the emerging needs, in order to offer innovative products and services, able to satisfy even the more or less latent needs of the consumer.

Companies are organising themselves to extract, from the data published on social networks, the most useful information to profile customers and to improve the products and services offered to meet their current and prospective needs. Initially the information was only obtained from the analysis of the published posts; more recently companies can obtain important indications on the consumption habits of customers also through the analysis of photographs published on social networks. In this way, companies are able to understand customers' consumption patterns of their products, monitor the most popular brands, and measure the level of identification of a consumer with a particular brand (Bezza and Giammario 2015).

In addition, social networks can also be a valuable help in acquiring information and understanding customer feedback about the strategies implemented by the competition, in order to identify possible reaction opportunities.

However, the rapid adoption of social media that accompanies and in some way encourages radical changes in customer behaviour also presents new risks for insurance companies.

The proliferation of both off-board platforms (public platform), such as Facebook, Twitter, Foursquare, YouTube, and others, and on-board platforms (brand-owned platform) has joined the expansion of direct digital channels such as mobile telephony, Internet, and interactive TV. This expansion leads to the fragmentation of contact points with customers, since new channels continuously emerge, making it more difficult to establish a lasting relationship. This fragmentation and the speed of communications make a brand likely be damaged in a very short time; in fact, there is no possibility to control the external social channels where the companies, after entering their content, can hardly counter or replace what is spread in the comments on the web.

It follows that the evolution that we are witnessing on the front of digitisation requires financial intermediaries to develop an integrated communication strategy, which allows them to reach the customer through social channels, in order to (1) better understand customer preferences; (2) create new needs and anticipate the needs of customers; (3) increase the level of satisfaction and, therefore, loyalty; and (4) monitor and oversee reputational dynamics that could otherwise escape control.

Different phases can be distinguished through which the development of a social media strategy is articulated (Viale and Souche 2012).

In an early step, publicly available social information is analysed primarily to identify the current filling of the customer and, secondly, to identify the influencers that can affect the customer's opinion and influence the decision-making process. In order to effectively conduct these activities, insurance companies identify and employ analysis tools and social media monitoring services in real time, 24 hours a day, seven days a week, in all geographic areas where the insurance company operates.

At a later stage, these activities are accompanied by activities aimed at engaging with customers in a systematic fashion, trying to establish and consolidate the presence on both on-board platforms and off-board platforms.

The company selects which social channels to supervise and which strategies to adopt, on the basis of the objectives set and the target customer. The initiatives can concern the use of the channel for the mere sharing of information on the products offered, but also for the improvement and support of the customer service or to guarantee the quick formulation of quotes. For example, by accessing the company's Facebook page and clicking on the "Buy Now" action button, users will be redirected to the company's website, where, by simply inserting the vehicle's license plate and the date of birth of the owner, they can get a quote for the car policy.

In the insurance sector, consumers often do not have adequate technical skills and struggle to trust insurance companies completely. The use of social media to clarify the doubts of users is useful for spreading the knowledge of the brand and trying to make the world of insurance more accessible to the consumer.

Some companies choose a very defined profile, focusing on a high interaction with the customer and user, varying the multimedia content, and directly offering special promotions exclusively dedicated to the social network customers. Other companies seem to have adopted a more balanced line, with a moderate level of interaction and a good balance between the presence of insurance issues and non-insurance-related ones.

The development of on-board social media is typically based on best-of-breed platforms that include, but are not limited to, blogs, reviews, forums, and sites of Q&A.

On the other hand, insurance companies can be present on off-board social media through the creation of pages or groups related to their brand on consolidated channels such as Facebook, Twitter, Google, or LinkedIn.

Different social platforms will be used, with respect to the different objectives to be achieved. For example, Twitter can be usefully adopted for customer service activities, while Facebook can be exploited not only to respond to customer requests, but also to strengthen the relationships with the community through advertisements, surveys, and the organisation of events. It is crucial to manage the interactions, in order to identify the factors that affect the reputation of the brand, by establishing an efficient auditing and reporting system, as regards both the on-board and the off-board activities (Kaur and Singh 2017). It is necessary in this regard to elaborate a careful governance structure, able to define the tasks of each company function involved (customer service, marketing, sales and recruiting team) on the different social platforms concerned.

It is also crucial to optimise the use of social channels according to a customer relationship management programme, as is already the case for traditional marketing channels. The optimised use of social media enables companies to extend the scope and flexibility of marketing campaigns aimed also at niche consumer groups, including those with specific interests or special needs.

5   CONCLUSIONS

The traditional contrast between the agency channel and the direct channel will have to be tackled in a different way in the years to come, leveraging a greater strategic awareness of technological and computer innovation. But social media are not additional sales channels: they are rather like a huge opportunity to increase corporate visibility, to innovate the relationship with the market along all phases of the customer lifecycle, from the estimate phase to post-sales management, and also to innovate the way in which functions within the enterprise are carried out.

New digital channels are considered increasingly important and preferable when it comes to comparing products, accessing contract information, comparing prices, and quickly enabling simple service requests. The 2.0 Web is therefore destined to deeply influence the organisational models of companies and agencies, both on the internal side and in the relationship with customers and with the realities outside the company.

However, the use of social media itself does not constitute the real change, nor can it give satisfactory results if it is freed from a logic of coherence with the values of the company.

The insurers who will take the greatest advantage are those who are able to integrate the use of social media within a global communicational approach, which includes all the media available to the company. It is necessary to connect social media to organisational processes (production and distribution) and ensure that they do not become the target, but the tool. Social media are a new reality that is still largely unexplored. In order for them to become business opportunities, it is necessary to bring about a cultural paradigm change, which can be translated into a new way of understanding the relationship between company, channels, and customers. Once the integration phase is complete, the real challenge will be to elaborate and correctly interpret the large mass of information that comes from the web and is useful for the development of the insurance business.

References

Badoc, M. (1986). *Marketing pour le banque et l'assurance europeennes*. Paris: Les editions d'organisation.

Baier, M. (1985). *Elements of Direct Marketing*. New York: McGraw Hill.

Bezza, Y., & Giammario, F. (2015, July). *Assicurazioni e social media*. La presenza social delle compagnie e l'evoluzione del rapporto con il cliente, KPMG.

Bird, D. (2007). *Common-Sense Direct & Digital Marketing* (5th ed.). London: Kogan Page.

Bloemer, J., de Ruyter, K., & Wetzels, M. (2009). Linking Perceived Service Quality and Service Loyalty: A Multi-Dimensional Perspective. *European Journal of Marketing, 33*, 1082–1106.

Boulding, W., Kalra, A., Staelin, R., & Zeithaml, V. (1993). A Dynamic Process Model of Service. *Journal of Marketing Research, 30*(1), 7–27.

Brasini, S., & Tassinari, G. (2014). Dalla Customer Satisfaction Alla Customer Loyalty. *Statistica Applicata, 16*(4), 443–468.

Brian, T., & Housden, M. (2017). *Direct and Digital Marketing in Practice*. London: Bloomsbury Publishing.

Brioschi, E. T. (1990). *la comunicazione d'azienda: caratteri fondamentali e problematiche di gestione*. Milano: Vita e Pensiero.

Butcher, K., Sparkes, B., & O'Callaghan, F. (2001). Evaluative and Relational Influences on Service Loyalty. *International Journal of Service Industry Management, 12*(4), 310–327.

Capgemini. (2014, April). World Insurance Report 2014: Leading with Digital for Better Customer Experience.

Cummins, J. D., & Venard, B. (2007). *Handbook of International Insurance: Between Global Dynamics and Local Contingiences*. Cham: Springer.

David Shepard Associates. (1995). *The New Direct Marketing, How to Implement a Profit-Driven Database Marketing Strategy Inc* (2nd ed.). Huntersville: Irwin.

De Ruyter, K., Wetzels, M., & Bloemer, J. (1998). On the Relationship Between Perceived Service Quality, Service Loyalty and Switching Costs. *International Journal of Service Industry Management, 9*(5), 436–453.

Delanote, S., Leus, R., & Nobibon, F. T. (2013). Optimization of the Annual Planning of Targeted Offers in Direct Marketing. *Journal of the Operational Research Society, 64*(12), 1770–1779.

Donnelly, J., Berry, L. L., & Thompson, T. W. (1985). *Marketing Financial Services: A Strategic Vision.* Huntersville: Dow Jones, Irwin.

Dumm, R. E., & Hoyt, R. E. (2003). Insurance Distribution Channels: Markets in Transition. *Journal of Insurance Regulation, 22*(1), 27–47.

Eiglier, P., & Langeard, E. (2003). *Il marketing strategico nei servizi.* Milano: McGraw Hill.

Goodman, E. (2014). Design and Ethics in the Era of Big Data. *Interactions, 21*(3), 22–24.

Hänninen, N., & Karjaluoto, H. (2017). The Effect of Marketing Communication on Business Relationship Loyalty. *Marketing Intelligence & Planning, 35*(4), 458–472.

Holtz, H. (1986). *Direct Marketing.* New York: John Wiley & Sons.

Iacobucci, D. (Ed.). (1999). *Handbook of Services Marketing and Management.* Thousand Oaks: Sage Publications.

Insurance Europe. (2017, May). Annual Report 2016–2017. Brussels.

Javalgi, R. G., & Moberg, C. R. (1997). Service Loyalty: Implications for Service Providers. *Journal of Services Marketing, 11*(3), 165–179.

Kaur, R., & Singh, G. (2017). Internet Marketing: The New Era of Innovation in E-Commerce. *International Journal of Scientific Research in Computer Science, Engineering and Information Technology, 2*(1), 253–258.

Kotler, P. (2001). *A Framework for Marketing Management.* Upper Saddle River: Prentice Hall.

Kotler, P., & Pfoertsch, W. (2011). *Ingredient Branding: Making the Invisible Visible.* New York: Springer.

Kroll, K. (2016). Direct Marketing of Personal Insurance. Gen Re. *Risk Insights,* No. 6.

Mela, C. F., Gupta, S., & Lehmann, D. R. (1997). The Long-Term Impact of Promotion and Advertising on Consumer Brand Choice. *Journal of Marketing Research, 34,* 248–261.

Nash, E. L. (Ed.). (1995). *Direct Marketing: Strategy, Planning and Execution* (3rd ed.). New York: McGraw-Hill.

Normann, R. (2001). *Service Management: Strategy and Leadership in Service Business* (3rd ed.). New York: Wiley & Sons.

Oliver, R. L. (1999). Whence Consumer Loyalty? *Journal of Marketing, 63,* 33–44.

Patterson, P. G., & Ward, T. (2000). Relationship Marketing and Management. In T. Swartz, H. Peck, A. Payne, M. Christopher, & M. Clark (Eds.), *Relationship Marketing: Strategy and Implementation.* Oxford: Butterworth-Heinemann.

Peng, O. K., & Mercer, A. (1998). The Direct Marketing of Insurance. *European Journal of Operational Research, 109*(3), 541–549.

Reicheld, F. F. (1996). *The Loyalty Effect.* Boston: Harvard Business School Press.

Roberts, M. (1989). *Direct Marketing Management.* London: Pearson.

Roman, E. (1995). *Integrated Direct Marketing.* New York: McGraw Hill.

Rowley, J. (2005). The Four Cs of Customer Loyalty. *Marketing Intelligence & Planning, 23*(6), 574–581.

Rust, R. T., & Zahorik, A. J. (1993). Customer Satisfaction, Customer Retention, and Market Share. *Journal of Retailing, 69*(2), 193–216.

Schwarz, G., Naujoks, H., Goossens, C., Whelan, D., Schwedel, A., & Singh, H. (2014). *Customer Loyalty and the Digital Transformation in P&C and Life Insurance.* Boston: Bain & Company.

Su Chen, M., & Lai, G. C. (2010). Distribution Systems, Loyalty and Performance. *International Journal of Retail & Distribution Management, 38*(9), 698–718.

Thornton, J., & White, L. (2001). Customer Orientations and Usage of Financial Distribution Channels. *Journal of Services Marketing, 15*(3), 168–185.

Viale, E., & Souche, C. (2012). *Insurers and Social Media.* Accenture.

Wang, Z., Singh, S. N., Li, Y. J., Mishra, S., Ambrose, M., & Biernat, M. (2017). Effects of Employees' Positive Affective Displays on Customer Loyalty Intentions: An Emotions-as-Social Information Perspective. *Acad Manage Journal, 60*(1), 109–129.

CHAPTER 6

Survey on the Digitised Insurance Distribution in Europe and USA

Abstract The chapter aims to analyse a sample of insurance services distribution websites located in Europe and USA. The description and the analysis of the models have been carried out from a potential customer's perspective, who visits the operators' websites in order to know the various insurance alternatives by surfing the internet; in this way, he is confident in the sole information available on the website he visited, as the relationship with the insurance company originates and develops exclusively or mainly through the web. The chapter ends with an analysis model which, still from a potential customer's perspective, permits the detection of the automated distribution distinctive traits, that is, accessibility, transparency, and quality of the offer, by basing the judgement on the objective elements the customer can gather through the digital channel.

Keywords Digitised insurance distribution • Insurance on line channels analysis • Customer experience

1 INTRODUCTION

Insurance distribution channels have evolved over the years in response to changes in customer behaviour and technological developments.

The distribution, once represented mainly by the agency network, has progressively evolved with new channels such as brokers, financial promoters

© The Author(s) 2018   99
A. Cappiello, *Technology and the Insurance Industry*,
https://doi.org/10.1007/978-3-319-74712-5_6

(for life and annuity products), banking channels, and direct sales, thanks to the application of digital technology that is becoming widespread.

Internet and social media penetration have grown significantly, and an increasing share of online activity is executed via mobile phones. If traditional intermediaries remain the main protagonists of the distribution in many segments, it is possible to assume that technological innovation will shortly play a vital role in this area, especially for what concerns easily standardised and low-consulting content products.

2   The Insurance Distribution Channels in the Main Industrialised Countries

For the moment, the widespread diffusion of digital technology has not revolutionised the role of traditional intermediaries within the entire insurance value chain. These, in fact, are still the leading operators in the global insurance distribution.

Agents, brokers, and other intermediaries such as banks account for a relatively stable share of around 60–70% of premiums in most insurance markets (Bain 2014; Swiss Re 2017; Power 2017).

Online sales of insurance remain relatively small in many countries, both compared with other distribution channels and e-commerce penetration in other sectors. In the European Union, for example, e-commerce sales by non-financial firms amounted to 16% of aggregate turnover in 2015 (up from 12% in 2008), and for some activities such as booking accommodation, the share of Internet sales is over 25%. This compares with an average share of direct online insurance of probably less than 5% (Eurostat 2017).

The causes of the limited spread of e-commerce insurance are manifold. In emerging countries, this can be due to the lack of adequate technologies; in advanced economies, on the other hand, insurers can be poorly prepared and not inclined to use digital technologies for the distribution of their services. An example thereof are small, community-based insurers in North and South America that, in particular, have a high affinity with traditional agent/broker distribution, perhaps linked to budget constraints on the necessary IT upgrades for digital distribution as well as potential worries about channel conflict. The diffusion of new technology into some wholesale insurance markets also remains patchy and here too, manual processes still dominate (Swiss Re 2016).

Several behavioural, institutional, and cultural factors also support the persistent role of insurance intermediaries. For example, some consumers may prefer to receive personal advice in purchasing an insurance product instead of searching and purchasing directly on the web. This is especially for products that have a high level of complexity. Similarly, buying insurance for considerable, complex commercial risks continues to be mostly done via brokers, who are often crucial in evaluating companies' risks and in matching their needs with an optimal provider (Eurostat 2016; Capgemini/Efma 2017).

The role of intermediaries in the sale of life insurance remains especially essential, probably reflecting the more complex nature of many of these products and the value prospective insureds attach to intermediaries' advice.

Among the largest life insurance markets, most products were sold via Bancassurance in Italy (80% of gross written premiums) and France (60%), while in the UK and Germany most life products were sold by agents and brokers (80% in both cases).

On the contrary, there is a more significant diffusion with digitised technologies for what concerns the distribution of standardised products of the non-life insurance.

In both large and small markets, non-life insurance policies are mainly distributed through agents and brokers. Agents predominate in Slovakia (80%), Italy (80%), Turkey (70%), Slovenia (60%), Germany (55%), and Portugal (54%). Meanwhile, brokers account for 60% of non-life premiums in Belgium and 50% in Bulgaria (Insurance Europe 2017; III 2017).

Even on the US market, the main channel is still the agency, which has not seen substantial decreases in its sale volumes, despite all the technological developments (Marsh 2016).

In this regard, the forecasts of experts indicate that the distribution of standardised products such as motor and household insurance will be the most affected area by the new technologies in the field of insurance (GSM Association 2015; Willis Towers Watson/Mergermarket 2017).

The digital distribution solutions of the insurance products will increasingly expand as the reliability and potential of the Internet will grow. Likewise, as insurers themselves become more experienced with digital technology, they will be able to offer customers a complete online purchase experience. This is especially likely if technological innovations facilitate usage-based insurance where consumers need to review, organise, and purchase insurance for particular activities at specific times (Bcg and Nice 2016; Swiss Re 2017).

If in the past the consumers turned to their agent or broker for all insurance needs, they are now more self-directed and use several tools to seek information, research, and purchase the insurance solutions they consider the best. Consumers are embracing innovation in financial services, mainly where it makes their interaction more convenient and improves communication. They want new products and services that respond to their needs and the added convenience of interacting with their insurers when, how, and where they want. No longer an annual transaction, the consumer/insurer relationship becomes more of a day-to-day experience (Christensen et al. 2015; McKinsey 2016).

Although the use of traditional channels remains predominant, it is still possible to detect that the digital technology is having a significant impact on the whole distribution process: that is, both regarding how products and services are delivered and more generally how companies interact with their customers (Swiss Re Institute 2017). This has been driven not only by customer expectations and a desire to increase their convenience and control, but also by the need to reduce distribution costs. Key features of this trend have been a renewed focus on the contact centre, the redirection of agents to handle more complex service transactions, and a continued shift towards a full self-service capability (Accenture 2015).

New digital offerings may merely provide alternative communication channels, such as email, website live chats, social media, online forums, or may make choosing or buying insurance more efficient, by using website self-service or mobile app (Nice-Bcg 2016).

The customer relationship will change even more in the future, because generational effects may also be significant in fostering Internet sales, with surveys indicating that younger policyholders are more likely to embrace new distribution channels (Yu and Portera 2015). This suggests that as the younger cohorts age and buy more insurance, online sales will likely increase.

While the agent–broker model may suit existing customers, new generations of insurance buyers will demand omni-channel, multitouch distribution (Capgemini/Efma 2017).

If this trend among millennials (born in 1980 to 1996) continues to grow, it could substantially change the way insurance companies interact with customers in the coming years.

However, millennials are the least satisfied of all generations with the online experience, and this can be a reason why this tech-driven age group has an overall low engagement with their primary insurer. For insurance company leaders, this means that improving the interactions with customers online is a smart investment towards building strong relationships within this future mainstream customer base.

3   The Choice of the Survey Sample

In order to detect the distinctive traits of the online distribution of insurance services, with particular regard to the aspects that most affect the insurer/customer relationship, a sample of insurance services distribution websites located in Europe and North America will be analysed; the latter market is in fact more exposed to the competitive pressure of the InsurTech. On the other hand, the European market is characterised by a marked lack of homogeneity as compared to the American reality, since it is much more diversified in the diffusion and use of digitised technologies for insurance distribution; among the scope of the latter there are several countries, such as Italy, which are still characterised by a significant presence of the insurance distribution through agencies and brokers.

The sample takes into consideration ten North American companies, and twenty five European companies, chosen among the leading insurance providers, which are referred to by the price comparison websites present in the geographical areas considered. For the European area, the selection was made to take into account the mentioned dimensional, managerial, and market differences existing between the different countries. Also, for each geographic area, we have only taken into account those companies that have a website and a personal area through which the customer, after having registered with a username and a personal password, can view and manage the policies subscribed (see Table 6.7 in Appendix).

The sample has included both traditional insurance companies that are approaching the digital world and direct insurance companies. In particular, we have identified companies with a so-called "pure" degree of automation, that is, that allow the customer to subscribe and buy insurance policies directly online, through their website, and companies with a "hybrid" degree of automation, that allow for the quote and the management of policies through the personal area on the web, but then refer to the direct contact with an agent for the purchase (Table 6.1).

Table 6.1 Components of the sample

	Europe	North America	Total
Insurance companies	25	10	35
Level of automation			
Pure	12	7	19
of which with live chat	2	2	4
Hybrid	13	3	16
of which with live chat	1	0	1
Property			
Insurance company	21	6	27
Mutual insurance	2	2	4
Bancassurance	1	0	1
Financial services company	1	2	3
The main service offered online			
Motor insurance	18	5	23
Travel insurance	2	0	2
Life insurance	1	3	4
Health insurance	2	1	3
Supplementary health insurance	1	1	2
Legal protection	1	0	1

The survey aims to test the characteristics of accessibility, transparency, and quality of the offer for the sample examined (Pia 2017), given the situation of a consumer who wants to subscribe an insurance policy for the first time on the basis of the sole information he or she can find on the Internet.

The observation was repeated at monthly intervals over the last year, to detect any change, either for the better or for the worse, occurring in the sample.

Table 6.2 shows that American companies have the highest level of automation since 70% of the companies considered give their customers the possibility to subscribe and buy their policies directly online, while in Europe, the percentage drops to 48%.

As far as the products offered are concerned, it is to note that in Europe online services are mainly in the area of motor insurance (72% of the companies examined), while American companies are geared towards life and health insurance in a greater proportion than Europe (50% of American companies as compared to 16% of European companies).

This difference is motivated by the fact that, in the USA, unlike in European countries, where healthcare is a universal right guaranteed by

Table 6.2 Main evident features from the website of the sample companies

Features	Europe	North America	Sample
Personalisation	44%	60%	48.57%
Economical features	40%	50%	42.85%
Professionalism	40%	20%	34.2%
Customer support	28%	20%	25.7%
Easy access	24%	20%	22.8%

law, the health system is based on criteria of an essentially private nature. While in Europe access to healthcare is guaranteed to all citizens (although there are substantial differences from country to country), regardless of the wealth and income, citizens in America have to cope with medical expenses autonomously and/or by using an insurance coverage. In this respect, there are federal health programmes, whose purpose is to help individuals or families with low income to bear health insurance costs, covering a more or less relevant part of it. Moreover, for employers of companies with more than 50 employees, there is an obligation to contribute to the costs of purchasing health policies for their workers. Estimates indicate that today about 87% of Americans have signed a private insurance policy for medical expenses, as there are penalties for citizens who have not taken out an insurance policy. In light of the above, it is therefore clear why health policies are more widespread in the USA than in Europe and why there are more American insurance companies that focus primarily on this service.

4   THE CUSTOMER RELATIONSHIP

Considering the visits that had been made on the selected sample websites, we can identify the characteristics of the offer that are highlighted more by the insurers in order to leverage the customers and convince them to take up their proposal. On average, every company emphasises at least two features of its business, listed in Table 6.2, in order of frequency.

As we can see, the feature on which insurance companies focus more, irrespective of the geographical area to which they belong, is the personalisation of the service. In the face of the high standardisation of products distributed online, the objective is to recover a relational and personalised approach with customers, communicating that it is possible to find appropriate solutions for their needs, even without having to recur to a *face-to-face* consulting service with an insurance agent.

The second highlighted feature, by order of importance, is the economic convenience, although this seems to be much more important for American companies (50% as compared to 40% of European companies), for which it is placed second, probably because of the fierce competitiveness of the American insurance market, characterised by a proportionally much higher number of insurers than in Europe.

In third place, we find the reference to professionalism, which seems to play an important role especially for European insurers (40% as compared to 20% of American companies) because many of these are placed in the first positions of the world ranking.

The fourth feature that insurance companies tend to emphasise is the offer of customer support, which is more or less of the same importance for both the geographical areas considered. Underlining this aspect, the company intends to reassure even the most inexperienced users about the availability of an assistance service, which is able to clarify all their doubts, assist them in choosing the most suitable product for their needs, and support and guide them in the subscription of the policy or in the event of a claim.

In the last place, we can find the ease with which the customer can access the service, particularly referred to one of the leading advantages offered by the automation of insurance services, given by the ease with which users can view the status of their policies or the progress of claims, regardless of the time or location.

The assessment of the quality of the digitisation level achieved by the online distribution of the services of the sample companies examined was carried out considering the profiles of *accessibility, transparency,* and *quality* of the service, which were in their turn evaluated through indicators; a specific score (from −1 to +5) has been attributed to these indicators according to the extent of their contribution to qualify the condition to which they are associated, as we can see from the analysis of their websites.

The automation of insurance services offers customers the advantage of being able to access their policy at any time and anywhere in the world, thus allowing them to make payments, to denounce claims, to request assistance even without the direct contact with an agent, provided they have a computer, a smartphone, or any device that allows them to connect to the Internet.

Table 6.3 Accessibility to the service (maximum score = 16)

Absence of barriers	Weight
International presence	1 point
Existence of physical agencies	1 point
Interaction	
Intuitive navigation	1 point
Online subscription	2 points
Live chat	1 point
Email	1 point
Phone	1 point
App	1 point
Social pages	1 point
Understanding	
Highlighting the service	2 points
Easy language	1 point
FAQs	2 points
Number FAQ ≥ 20	1 point

In order to test the level of accessibility to the services offered by the companies analysed, we have taken into account the absence of barriers (geographical and possibly physical), the level of interaction with the user, and the level of intelligibility of the reported information as you can see in Table 6.3.

The level of interaction is due to the ease with which even the most inexperienced users, without any particular computer skills, can access the service. In fact, the structure of the web page should allow the user to navigate efficiently, to search for useful information in order to really understand the offer, to obtain the relevant information, and to return to a point previous (ASIC 2016). The level of interaction is also evaluated on the basis of the channels and devices (live chat, email, phone, and mobile app) used to contact the company itself or the customer service centre. The more numerous the points of contact are, the more convenient the service becomes, ensuring the accessibility at all times and in every place (Visciola 2006).

As far as the understanding of the contents is concerned, it is of paramount importance that the information is exposed in a clear way and in an easy language. To further improve the understanding of the proposals, it is useful to have a section containing the most frequently asked questions (FAQs) with their answers. As the FAQs generally reflect the real difficulties experienced by customers, reading them can be very useful in order to clarify doubts or avoid misunderstandings.

Table 6.4 Transparency of the service (maximum score = 11)

Company information	Weight
History and main property statements	1 point
Staff with photos	1 point
Permission	1 point
Conditions	
Indication of deductible or percentage excess	1 point
Indication of exclusions or compensations	1 point
Online claim advice mode	2 points
Detailed information booklet	2 points
Costs clearly indicated	2 points

The second aspect considered relates to the level of information transparency (Table 6.4).

The insurance company, regardless of how it delivers the service, is obliged to provide users with truthful, transparent, and sufficiently detailed information so that potential customers can understand the business model and the organisational structure of the company, appreciate its equity strength, understand the characteristics of the offer, as well as the specific contractual conditions of the individual products and services, so that customers can make informed decisions.

Companies that offer the online subscription service must pay particular attention to the amount and quality of available information, including those required by law, which must be readily available on the web page, considering the lack of a personal relationship with the customer. The information released, besides being of high quality, must also be readily understood, so as to be able to satisfy even customers with poor financial culture and limit the risks of litigation.

Finally, the third aspect analysed is related to the quality of the service (Table 6.5).

The conditions that contribute to express the quality of the offer refer to the user experience that the website can offer to the customer, to the variety of services offered, and also to those circumstances that reduce the quality perceived by users, such as the presence of any problems while browsing (Hassenzahl and Tractinsky 2011; Sward and Macarthur 2007).

The quality of the service has also been analysed by attributing a score to the different indicators, even if, in the specific case, a higher level of subjectivity affects the score, since it was, in some cases, a real assessment, instead of ascertaining whether or not particular items were present.

Table 6.5 Quality of the service (maximum score = 22)

User experience	Weight
Website availability	From 0 to 5 points
Video/interactivity	1 point
Graphics	From 0 to 5 points
Updated social pages	From 0 to 5 points
Possibility to leave reviews	1 point
Business model	
Variety of services offered	From 0 to 5 points
Problems encountered	
Page refresh issues	1 point
Page not found	1 point

5   Results Analysis and Assessment

The final score reported by each company belonging to the sample analysed is the result of the summation of the scores assigned to the qualitative evaluation of the profiles of accessibility, transparency, and quality of service (Graph 6.1).

Both the maximum registered score, equal to 44 points, and the minimum, of 27, are referred to a European company. In fact, although European companies have registered, on average, a higher value (37.1 points) than the American ones (36 points), it is to note that the European subsample has a higher variability in the scores, with a higher standard deviation (equal to 4.09) compared to North America (2.05), which confirms a greater dispersion of the recorded values than the average value (Table 6.6).

This variability is explained by the fact that the European companies analysed reflect the features of heterogeneity that characterises the insurance distribution at European level where, alongside technologically evolved realities, there are traditional companies that have only recently implemented digital solutions. The latter, if we exclude the major European insurance groups, which approach the digital world more rigorously, show a certain backwardness compared to the American competitors, especially as far as the variety and quality of services offered online are concerned.

It seems useful to better study the breakdown of the final score through the analysis of its three components related to accessibility, transparency, and quality of service, as shown in Table 6.6.

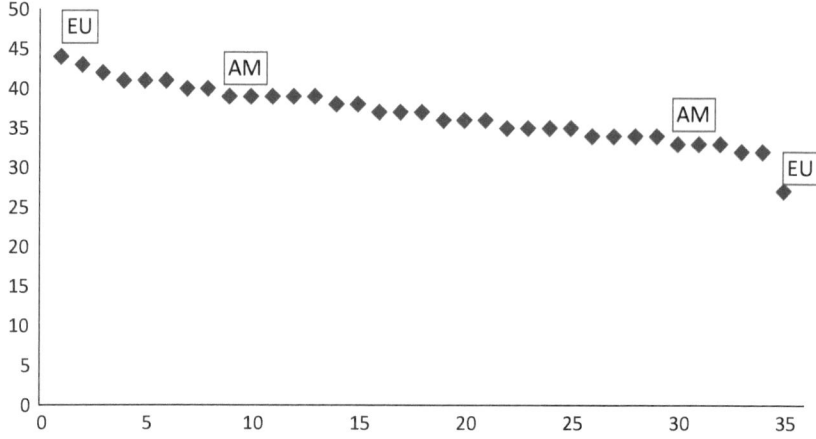

Graph 6.1   Total score reported by each company

Table 6.6   Survey results

	Accessibility to the service (Average value)	Transparency (Average value)	Quality of the service (Average value)	Total average value	Standard deviation
North America	13.4	4.6	18	36	2.05
Belgium	12	7.5	15	34.5	
France	13.4	8	18.4	39.8	
Germany	12	7.6	17.3	36.9	
Italy	12	8.2	16.7	36.9	
Luxembourg	12	4	16	32	
Netherlands	14	7.5	16.5	38	
United Kingdom	13	7.6	17.3	37.9	
Spain	11	7	16.6	34.6	
Switzerland	11.5	8	18	37.5	
Total Europe	12.4	7.6	17.1	37.1	4.09
Total sample	12.9	6.1	17.5		

Regarding the requirement of accessibility of the service, the investigated sample obtains an average score of 12.9, with a gap of one point between Europe and North America. The traditional companies, especially the European ones, show a certain level of backwardness compared to the American ones, which are technologically more advanced. It should be noted that, at the moment, only 54% of the companies considered offer the online subscription service, whereas the remainder refers to a direct contact with an agent, allowing the insureds just to monitor their policies and send off the claim reporting form via the website. It is to note, however, a reasonable level of interaction of the insurer/consumer contact is guaranteed, since the companies provide their customers, on average, with three ways to contact the company, confirming their particular carefulness in trying to provide constant customer support. The level of intelligibility of the digital content on the websites normally receives a good score: companies demonstrate a strong commitment in trying to adopt an accessible and understandable language even for the less-experienced ones in financial services. It should also be noted that 83% of the companies have developed a section of their website dedicated to FAQs, implementing a wide range of questions in order to emphasise the quality of their offer and clarify any doubts of the users.

Regarding the level of transparency, the companies analysed show, on average, a score equal to 6.10 out of a maximum of 11. This global result is penalised by the fact that it is not possible, in several cases, to determine the exact costs of the products and services offered before completing a process of quotes, reserved, for most companies, to users residing in the geographic area lying in the company's jurisdiction. It should also be noted that the burden of any fees or commissions due to the insurance undertaking is not expressed, thus not allowing immediate and transparent information on the price formation.

As regards the other transparency items, it was found that about 70% of companies disclose information about their history, their staff, the qualification of supervised subjects, the solvency situation, and information concerning the conditions of the products offered, such as the presence of any contract clauses such as deductibles or exclusions in general. The latter information, however, is missing in most American companies since it was not possible to access the informative dossiers of the various products offered, probably because they are reserved to users who ask for a quotation or who, being already customers, have a personal username and password.

This deficiency lowers the transparency score attributed to American companies which, in fact, is significantly lower than that of European countries, notably France, Italy, and Switzerland.

As regards the quality of the service offered, the sample companies show, on average, a score of 17.5 out of a maximum of 22 points. The American companies record the highest score (18 points), whereas the European ones are, on average, on a value of 17.1 points. In this respect, it should be pointed out that the score of the French, Swiss, German, and British companies, which is aligned to, if not even higher than, the American subsample, is penalised by the results of Belgium, Italy, the Netherlands, and Spain, which are on lower levels. The major European insurance groups, established globally, and the American companies have an articulated range of services and use more interactive websites, which can create higher user engagement. This apparently does not occur for some European companies whose sites, at present, have a poor interactivity and, in some cases, a very simple and somewhat colourless graphic. Finally, as regards the availability of the website, on average the companies examined show the maximum score of 5, since all the sites turn up among the first entries of the first page of the search engine, and there are no particular malfunctions; only 8% of the companies examined have problems related to the updating of the pages or the impossibility to open the contents of specific pages.

6   CONCLUSIONS

Although the survey is based on a statistically unrepresentative sample, it can be considered a valuable reference point for digitised distribution models that are currently being adopted in the analysed insurance markets.

The analysis shows that the American insurance companies have a higher average level of accessibility and quality of the services offered. However, they demonstrate shortcomings about the quality of information transparency. In fact, the simple surfing among the pages of the sites does not allow access to the information files of the products offered, in case you have not yet requested a quote or registered to the service. Compared to the European reality, regulated by strict rules on information transparency, this leads to a loss of informative immediacy and makes the intercompany comparison activity more difficult.

On the other hand, European companies are the ones that have obtained the best average score although, as already mentioned, there is a certain asymmetry between the different geographic areas of reference. The major European insurance groups show the highest score about international presence; they also have a specific website for each country in which they operate. The need to serve such a vast number of customers, in different geographic areas characterised by differentiated competitors, has facilitated, among other things, a more rigorous approach to the digitisation of every aspect of the distribution chain of these insurers.

Within the European subsample, however, the position of the Italian, Belgian, Spanish, and Luxembourg companies that were analysed stands out; although they present scores in line with the sample as regards transparency, they still show a certain backwardness as to the requirements of accessibility and quality of the service offered.

It can be said that, although the considerable effort made by the traditional insurance companies belonging to different geographic areas is evident, there are still ample margins of improvement. Despite the implementation of a personal area, reserved for every customer, within its website, many companies, at the moment, do not offer the online subscription service that, at least for the more standardised products, such as car insurance, could lead to significant benefits for both the customer and the company, including not having any place or time restrictions when it comes to signing a policy.

Another aspect that can be improved is the quality of the service, since several companies have little interactive sites, and a limited number of companies give the possibility to leave reviews and read those left by other users who have already experienced the service.

In all markets, insurance companies are accelerating the shift to a radically different distribution model, where digitalisation plays an increasingly important role in the majority of interactions, and agents' efforts are being refocused to add more value. Digital technology will have a significant impact on both the design of new products and the way they are delivered. It can be assumed that the partnership with the InsurTech start-up by the incumbent insurance can help to reconfigure the distribution model and the modalities of the interaction with the customer. The lack of physical proximity to the customer and the relational approach typical of traditional intermediation can be recovered through the continuous implementation of advanced technological solutions able to allow for a more accurate customer profiling and consequently an upgrade of the whole customer experience.

APPENDIX

Table 6.7   The survey sample

Country	Insurance companies
Belgium	P&V Assurances scrl; Partenamut
France	AXA; Groupama; Europ Assistance Group; Crédit Agricole S.A; April Group
Germany	Allianz SE; Europäische Reiseversicherung AG (ERV); ARAG SE
Italy	Genertel S.p.a; UnipolSai Assicurazioni S.p.a; Reale Mutua Assicurazioni; Amissima Assicurazioni S.p.a
Luxembourg	La Luxembourgeoise
Netherland	Aegon N.V; ONVZ
Spain	Mapfre S.A; LD-Nuez; Mutua Madrileña
Switzerland	Zurich Insurance Group AG; Helvetia Assurances Suisse
United Kingdom	Aviva plc; Direct Line; RSA Insurance Group plc
USA	Metropolitan Life Insurance Company (MetLife); American International Group, Inc. (AIG); State Farm Mutual Automobile Insurance Company; American Family Insurance; The Travelers Companies, Inc.; Prudential Financial, Inc.; Aflac Inc.; Allstate Corporation; Government Employees Insurance Company (GEICO)
Canada	Manulife Financial Corporation

REFERENCES

Accenture. (2015). Reimaging Insurance Distribution.

ASIC. (2016, March). Providing Digital Financial Product Advice to Retail Clients.

Bain. (2014). Customer Loyalty and the Digital Transformation in P&C and Life Insurance, Global Edition.

Boston Consulting Group (BCG) and NICE. (2016, May). Digital Technologies Raise the Stakes in Customer Service.

Capgemini/Efma. (2017). World Insurance Report 2017.

Christensen, C. M., Raynor, M. E., & McDonald, R. (2015, December). What Is Disruptive Innovation? *Harward Business Review.*

Eurostat. (2016). Survey on ICT Usage and E-Commerce in Enterprises.

Eurostat. (2017, March). Statistic Yearbook, Statistic Explained.

GSM Association. (2015). *Mobile Insurance.* Savings and Credit Report, London.

Hassenzahl, M., & Tractinsky, N. (2011). User Experience—A Research Agenda. *Behaviour & Information Technology, 25,* 91–97.

III—Insurance Information Institute. (2017). Facts + Statistics: Distribution Channels.

Insurance Europe. (2017, August). Insurance Data.

Marsh. (2016, May). United States Insurance Market Report 2016.

McKinsey. (2016). Digital and Auto Insurers Value at Stake Analysis.

NICE and BCG. (2016, October). Customer Survey Results.

Pia, P. (2017). *La consulenza finanziaria automatizzata*. Milano: FrancoAngeli.

Power, J. D. (2017, April). U.S. Insurance Shopping Study.

Sward, D., & Macarthur, G. (2007). Making User Experience a Business Strategy. In E. Law, A. Vermeeren, M. Hassenzahl, & M. Blythe (Eds.), *Toward a UX Manifesto*. Lancaster.

Swiss Re. (2016, July). Mutual Insurance in the 21th Century: Back to the Future? Sigma n.4.

Swiss Re. (2017, May). World Insurance in 2016: The China Growth Engine Steams Ahead, Sigma n.3.

Swiss Re Institute. (2017, June). Technology and Insurance: Themes and Challenges.

Visciola, M. (2006). *L'usabilità dei siti web*. Milano: Apogeo.

Willis Towers Watson/Mergermarket. (2017, February). Insurers Under Pressure to Go Digital.

Yu, D., & Portera, C. (2015, March 5). Insurance Companies Have a Big Problem with Millennials. *Business Journal, 3*, 84–91.

INDEX

© The Author(s) 2018
A. Cappiello, *Technology and the Insurance Industry*,
https://doi.org/10.1007/978-3-319-74712-5